Get the eBooks FREE!

(PDF, ePub, Kindle, and liveBook all included)

We believe that once you buy a book from us, you should be able to read it in any format we have available. To get electronic versions of this book at no additional cost to you, purchase and then register this book at the Manning website.

Go to https://www.manning.com/freebook and follow the instructions to complete your pBook registration.

That's it!
Thanks from Manning!

Deep Learning and the Game of Go

Deep Learning
and the Game of Go

MAX PUMPERLA
AND KEVIN FERGUSON

MANNING
Shelter Island

For online information and ordering of this and other Manning books, please visit
www.manning.com. The publisher offers discounts on this book when ordered in quantity.
For more information, please contact

    Special Sales Department
    Manning Publications Co.
    20 Baldwin Road
    PO Box 761
    Shelter Island, NY 11964
    Email: orders@manning.com

Manning Publications Co.
20 Baldwin Road
PO Box 761
Shelter Island, NY 11964

Development editor:	Jenny Stout
Technical development editor:	Charles Feduke
Review editor:	Ivan Martinović
Project editor:	Lori Weidert
Copyeditor:	Sharon Wilkey
Proofreader:	Michelle Melani
Technical proofreader:	Tanya Wilke
Typesetter:	Gordan Salinovic
Cover designer:	Marija Tudor

ISBN 9781617295324
Printed in the United States of America
1 2 3 4 5 6 7 8 9 10 – SP – 23 22 21 20 19 18

To Anne, it's all for you.

—Max

To Ian

—Kevin

brief contents

contents

foreword

For us, the members of the AlphaGo team, the AlphaGo story was the adventure of a lifetime. It began, as many great adventures do, with a small step—training a simple convolutional neural network on records of Go games played by strong human players. This led to pivotal breakthroughs in the recent development of machine learning, as well as a series of unforgettable events, including matches against the formidable Go professionals Fan Hui, Lee Sedol, and Ke Jie. We're proud to see the lasting impact of these matches on the way Go is played around the world, as well as their role in making more people aware of, and interested in, the field of artificial intelligence.

But why, you might ask, should we care about games? Just as children use games to learn about aspects of the real world, so researchers in machine learning use them to train artificial software agents. In this vein, the AlphaGo project is part of DeepMind's strategy to use games as simulated microcosms of the real world. This helps us study artificial intelligence and train learning agents with the goal of one day building general purpose learning systems capable of solving the world's most complex problems.

AlphaGo works in a way that is similar to the two modes of thinking that Nobel laureate Daniel Kahnemann describes in his book on human cognition, *Thinking Fast and Slow*. In the case of AlphaGo, the slow mode of thinking is carried out by a planning algorithm called *Monte Carlo Tree Search*, which plans from a given position by expanding the game tree that represents possible future moves and counter moves. But with roughly 10^170 (1 followed by 170 0s) many possible Go positions, searching through every sequence of a game proves impossible. To get around this and to reduce the size of the search space, we paired the Monte Carlo Tree Search with a *deep learning* component—two neural networks trained to estimate how likely each side is to win, and what the most promising moves are.

A later version, AlphaZero, uses principles of *reinforcement learning* to play entirely against itself, eliminating the need for any human training data. It learned from scratch the game of Go (as well as chess and shogi), often discovering (and later discarding) many strategies developed by human players over hundreds of years and creating many of its own unique strategies along the way.

Over the course of this book, Max Pumperla and Kevin Ferguson take you on this fascinating journey from AlphaGo through to its later extensions. By the end, you will not only understand how to implement an AlphaGo-style Go engine, but you will also have great practical understanding of some of the most important building blocks of modern AI algorithms: Monte Carlo Tree Search, deep learning, and reinforcement learning. The authors have carefully tied these topics together, using the game of Go as an exciting and accessible running example. As an aside, you will have learned the basics of one of the most beautiful and challenging games ever invented.

Furthermore, the book empowers you from the beginning to build a working Go bot, which develops over the course of the book, from making entirely random moves to becoming a sophisticated self-learning Go AI. The authors take you by the hand, providing both excellent explanations of the underlying concepts, as well as executable Python code. They do not hesitate to dive into the necessary details of topics like data formats, deployment, and cloud computing necessary for you to actually get your Go bot to work and play.

In summary, *Deep Learning and the Game of Go* is a highly readable and engaging introduction to modern artificial intelligence and machine learning. It succeeds in taking what has been described as one of the most exciting milestones in artificial intelligence and transforming it into an enjoyable first course in the subject. Any reader who follows this path will be equipped to understand and build modern AI systems, with possible applications in all those situations that require a combination of "fast" pattern matching and "slow" planning. That is, the thinking fast and slow required for basic cognition.

—THORE GRAEPEL, RESEARCH SCIENTIST, DEEPMIND,
ON BEHALF OF THE ALPHAGO TEAM AT DEEPMIND

preface

When AlphaGo hit the news in early 2016, we were extremely excited about this groundbreaking advancement in computer Go. At the time, it was largely conjectured that human-level artificial intelligence for the game of Go was at least 10 years in the future. We followed the games meticulously and didn't shy away from waking up early or staying up late to watch the broadcasted games live. Indeed, we had good company—millions of people around the globe were captivated by the games against Fan Hui, Lee Sedol, and later Ke Jie and others.

Shortly after the emergence of AlphaGo, we picked up work on a little open source library we coined BetaGo (see http://github.com/maxpumperla/betago), to see if we could implement some of the core mechanisms running AlphaGo ourselves. The idea of BetaGo was to illustrate some of the techniques behind AlphaGo for interested developers. While we were realistic enough to accept that we didn't have the resources (time, computing power, or intelligence) to compete with DeepMind's incredible achievement, it has been a lot of fun to create our own Go bot.

Since then, we've had the privilege to speak about computer Go on quite a few occasions. As we are both long-term Go enthusiasts *and* machine learning practitioners, it was at times easy to forget just how little the general public picked up from the events we followed so closely. In fact, it was a little ironic to see that while millions watched the games, at least from our perspective in the western world, there seem to be essentially two disjointed groups:

- Those who understand and love the game of Go, but know little about machine learning.
- Those who understand and appreciate machine learning, but barely know the rules of Go.

To an outsider, both disciplines might seem equally opaque, complicated, and hard to master. While in the last years more and more software developers picked up machine learning and in particular *deep learning*, the game of Go remains largely unknown to many in the west. We think this is very unfortunate and it is our sincere hope that this book brings the above two groups closer together.

We strongly believe that the principles underpinning AlphaGo can be taught to a general software engineering audience in a practical manner. Enjoyment and understanding of Go comes from *playing it* and experimenting with it. It can be argued that the same holds true for machine learning, or any other discipline, for that matter.

If you share some of our enthusiasm for either Go or machine learning (hopefully both!) at the end of this book, we've done our job. If, on top of that, you know how to build and ship a Go bot and run your own experiments, many other interesting artificial intelligence applications will be accessible to you as well. Enjoy the ride!

acknowledgments

We'd like to acknowledge the whole team at Manning for making this book possible. In particular, we'd like to thank both of our tireless editors: Marina Michaels, for getting us the first 80% of the way there; and Jenny Stout, for getting us through the second 80%. Thanks also to our technical editor, Charles Feduke, and our technical proofreader, Tanya Wilke, for combing through all of our code.

We'd also like thank all the reviewers who provided valuable feedback: Aleksandr Erofeev, Alessandro Puzielli, Alex Orlandi, Burk Hufnagel, Craig S. Connell, Daniel Berecz, Denis Kreis, Domingo Salazar, Helmut Hauschild, James A. Hood, Jasba Simpson, Jean Lazarou, Martin Møller Skarbiniks Pedersen, Mathias Polligkeit, Nat Luengnaruemitchai, Pierluigi Riti, Sam De Coster, Sean Lindsay, Tyler Kowallis, and Ursin Stauss.

Thanks also go to everyone who experimented with or contributed to our BetaGo project, especially Elliot Gerchak and Christopher Malon.

Finally, thanks to everyone who ever tried to teach a computer to play Go and shared their research.

I would like to thank Carly, for her patience and support; and Dad and Gillian, for teaching me how to write.

—Kevin Ferguson

Special thanks to Kevin for bringing this home, Andreas for many fruitful discussions, and Anne for her constant support.

—Max Pumperla

about this book

Deep Learning and the Game of Go is intended to introduce modern machine learning by walking through a practical and fun example: building an AI that plays Go. By the end of chapter 3, you can make a working Go-playing program, although it will be laughably weak at that point. From there, each chapter introduces a new way to improve your bot's AI; you can learn about the strengths and limitations of each technique by experimenting. It all culminates in the final chapters, where we show how AlphaGo and AlphaGo Zero integrate all the techniques into incredibly powerful AIs.

Who should read this book

This book is for software developers who want to start experimenting with machine learning, and who prefer a practical approach over a mathematical approach. We assume you have a working knowledge of Python, although you could implement the same algorithms in any modern language. We don't assume you know anything about Go; if you prefer chess or some similar game, you can adapt most of the techniques to your favorite game. If you *are* a Go player, you should have a blast watching your bot learn to play. We certainly did!

Roadmap

The book has three parts that cover 14 chapters and 5 appendices. *Part I: Foundations* introduces the major concepts for the rest of the book.

- Chapter 1, *Towards deep learning*, gives a lightweight and high-level overview of the discipline's artificial intelligence, machine learning, and deep learning. We explain how they interrelate and what you can and cannot do with techniques from these fields.
- Chapter 2, *Go as a machine learning problem*, introduces the rules of Go and explains what we can hope to teach a computer playing the game.
- Chapter 3, *Implementing your first Go bot*, is the chapter in which we implement the Go board, placing stones and playing full games in Python. At the end of this chapter you can program the weakest Go AI possible.

Part II: Machine learning and game AI presents the technical and methodological foundations to create a strong go AI. In particular, we will introduce three pillars, or techniques, that AlphaGo uses very effectively: *tree search, neural networks,* and *reinforcement learning.*

Tree search

- Chapter 4, *Playing games with tree search*, gives an overview of algorithms that search and evaluate sequences of game play. We start with the simple brute-force minimax search, then build up to advanced algorithms such as alpha-beta pruning and Monte Carlo search.

Neural networks

- Chapter 5, *Getting started with neural networks*, gives a practical introduction into the topic of artificial neural networks. You will learn to predict handwritten digits by implementing a neural network from scratch in Python.
- Chapter 6, *Designing a neural network for Go data*, explains how Go data shares traits similar to image data and introduces convolutional neural networks for move prediction. In this chapter we start using the popular deep learning library Keras to build our models.
- Chapter 7, *Learning from data: a deep learning bot*, we apply the practical knowledge acquired in the preceding two chapters to build a Go bot powered by deep neural networks. We train this bot on actual game data from strong amateur games and indicate the limitations of this approach.
- Chapter 8, *Deploying bots in the wild*, will get you started with serving your bots so that human opponents can play against it through a user interface. You will also learn how to let your bots play against other bots, both locally and on a Go server.

Reinforcement learning

- Chapter 9, *Learning by practice: reinforcement learning*, covers the very basics of reinforcement learning and how we can use it for self-play in Go.

- Chapter 10, *Reinforcement learning with policy gradients*, carefully introduces policy gradients, a vital method in improving move predictions from chapter 7.
- Chapter 11, *Reinforcement learning with value methods*, shows how to evaluate board positions with so-called value methods, a powerful tool when combined with tree search from chapter 4.
- Chapter 12, *Reinforcement learning with actor-critic methods*, introduces techniques to predict the long-term value of a given board position and a given next move, which will help us choose next moves efficiently.

Part III: Greater than the sum of its parts is the final part, in which all building blocks developed earlier culminate in an application that is close to what AlphaGo does.

- Chapter 13, *Alpha Go: Bringing it all together*, is both technically and mathematically the pinnacle of this book. We discuss how first training a neural network on Go data (chapters 5–7) and then proceeding with self-play (chapters 8–11), combined with a clever tree search approach (chapter 4) can create a super-human-level Go bot.
- Chapter 14, *AlphaGo Zero: Integrating tree search with reinforcement learning*, the last chapter of this book, describes the current state of the art in board game AI. We take a deep dive into the innovative combination of tree search and reinforcement learning that powers AlphaGo Zero.

In the appendices, we cover the following topics:

- Appendix A, *Mathematical foundations*, recaps some basics of linear algebra and calculus, and shows how to represent some linear algebra structures in the Python library NumPy.
- Appendix B, *The backpropagation algorithm*, explains the more math-heavy details of the learning procedure of most neural networks, which we use from chapter 5 onwards.
- Appendix C, *Go programs and servers*, provides some resources for readers who want to learn more about Go.
- Appendix D, *Training and deploying bots using Amazon Web Services*, is a quick guide to running your bot on an Amazon cloud server.
- Appendix E, *Submitting a bot to the Online Go Server (OGS)*, shows how to connect your bot to a popular Go server, where you can test it against players around the world.

The figure on the following page summarizes the chapter dependencies.

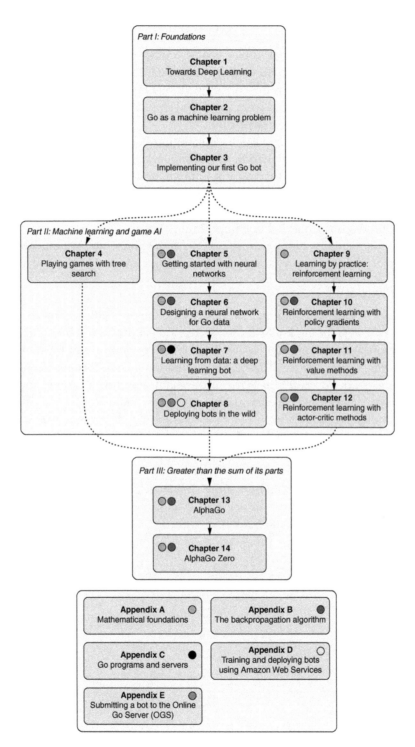

About the code

This book contains many examples of source code both in numbered listings and in line with normal text. In both cases, source code is formatted in a `fixed-width font like this` to separate it from ordinary text. Sometimes code is also **in bold** to highlight code that has changed from previous steps in the chapter, such as when a new feature adds to an existing line of code.

In many cases, the original source code has been reformatted; we've added line breaks and reworked indentation to accommodate the available page space in the book. In rare cases, even this was not enough, and listings include line-continuation markers (➥). Additionally, comments in the source code have often been removed from the listings when the code is described in the text. Code annotations accompany many of the listings, highlighting important concepts.

All code samples, along with some additional glue code, are available on GitHub at: https://github.com/maxpumperla/deep_learning_and_the_game_of_go.

Book forum

Purchase of *Deep Learning and the Game of Go* includes free access to a private web forum run by Manning Publications, where you can make comments about the book, ask technical questions, and receive help from the author and from other users. To access the forum, go to https://forums.manning.com/forums/deep-learning-and-the-game-of-go. You can also learn more about Manning's forums and the rules of conduct at https://forums.manning.com/forums/about.

Manning's commitment to our readers is to provide a venue where a meaningful dialogue between individual readers and between readers and the authors can take place. It is not a commitment to any specific amount of participation on the part of the authors, whose contribution to the forum remains voluntary (and unpaid). We suggest you try asking the authors some challenging questions lest their interest stray! The forum and the archives of previous discussions will be accessible from the publisher's website as long as the book is in print.

about the authors

MAX PUMPERLA is a Data Scientist and Engineer specializing in Deep Learning at the artificial intelligence company skymind.ai. He is the co-founder of the Deep Learning platform aetros.com.

KEVIN FERGUSON has 18 years of experience in distributed systems and data science. He is a data scientist at Honor, and has experience at companies such as Google and Meebo. Together, Max and Kevin are co-authors of betago, one of very few open source Go bots, developed in Python.

about the cover illustration

The figure on the cover of *Deep Learning and the Game of Go* is Emporer Montoku, who ruled Japan from 850 to 858. The portrait was done in watercolor on silk by an unknown artist. It was reproduced as part of "Emperors and Empresses of the Past" in the Japanese history journal *Bessatsu Rekishi Dokuhon* in 2006.

Figures like this one remind us vividly of the uniqueness and individuality of the world's towns and regions long ago. It was a time when the dress codes of two regions separated by a few dozen miles identified people uniquely as belonging to one or the other.

Dress codes have changed since then, and the diversity by region, so rich at the time, has faded away. It's now often hard to tell the inhabitant of one continent from another. Perhaps we've traded a cultural and visual diversity for a more varied personal life—or a more varied and interesting intellectual and technical life. We at Manning celebrate the inventiveness, the initiative, and the fun of the computer business with book covers based on the rich diversity of regional life centuries ago.

Part 1

Foundations

What is machine learning? What is the game of Go, and why was it such an important milestone for game AI? How is teaching a computer to play Go different from teaching it to play chess or checkers?

In this part, we answer all those questions, and you'll build a flexible Go game logic library that will provide a foundation for the rest of the book.

Toward deep learning: a machine-learning introduction

This chapter covers:

- Machine learning and its differences from traditional programming
- Problems that can and can't be solved with machine learning
- Machine learning's relationship to artificial intelligence
- The structure of a machine-learning system
- Disciplines of machine learning

As long as computers have existed, programmers have been interested in *artificial intelligence* (AI): implementing human-like behavior on a computer. Games have long been a popular subject for AI researchers. During the personal computer era, AIs have overtaken humans at checkers, backgammon, chess, and almost all classic board games. But the ancient strategy game Go remained stubbornly out of reach for computers for decades. Then in 2016, Google DeepMind's AlphaGo AI challenged 14-time world champion Lee Sedol and won four out of five games. The next revision of AlphaGo was completely out of reach for human players: it won 60 straight games, taking down just about every notable Go player in the process.

AlphaGo's breakthrough was enhancing classical AI algorithms with machine learning. More specifically, AlphaGo used modern techniques known as *deep learning*— algorithms that can organize raw data into useful layers of abstraction. These techniques aren't limited to games at all. You'll also find deep learning in applications for identifying images, understanding speech, translating natural languages, and guiding robots. Mastering the foundations of deep learning will equip you to understand how all these applications work.

Why write a whole book about computer Go? You might suspect that the authors are die-hard Go nuts—OK, guilty as charged. But the real reason to study Go, as opposed to chess or backgammon, is that a strong Go AI requires deep learning. A top-tier chess engine such as Stockfish is full of chess-specific logic; you need a certain amount of knowledge about the game to write something like that. With deep learning, you can teach a computer to imitate strong Go players, even if you don't understand what they're doing. And that's a powerful technique that opens up all kinds of applications, both in games and in the real world.

Chess and checkers AIs are designed around reading out the game further and more accurately than human players can. There are two problems with applying this technique to Go. First, you can't read far ahead, because the game has too many moves to consider. Second, even if you could read ahead, you don't know how to evaluate whether the result is good. It turns out that deep learning is the key to unlocking both problems.

This book provides a practical introduction to deep learning by covering the techniques that powered AlphaGo. You don't need to study the game of Go in much detail to do this; instead, you'll look at the general principles of the way a machine can learn. This chapter introduces machine learning and the kinds of problems it can (and can't) solve. You'll work through examples that illustrate the major branches of machine learning, and see how deep learning has brought machine learning into new domains.

1.1 *What is machine learning?*

Consider the task of identifying a photo of a friend. This is effortless for most people, even if the photo is badly lit, or your friend got a haircut or is wearing a new shirt. But suppose you want to program a computer to do the same thing. Where would you even begin? This is the kind of problem that machine learning can solve.

Traditionally, computer programming is about applying clear rules to structured data. A human developer programs a computer to execute a set of instructions on data, and out comes the desired result, as shown in figure 1.1. Think of a tax form: every box has a well-defined meaning, and detailed rules indicate how to make various calculations from them. Depending on where you live, these rules may be extremely

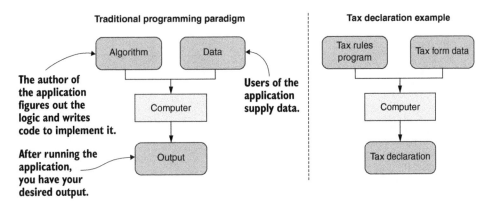

Figure 1.1 The standard programming paradigm that most software developers are familiar with. The developer identifies the algorithm and implements the code; the users supply the data.

complicated. It's easy for people to make a mistake here, but this is exactly the kind of task that computer programs excel at.

In contrast to the traditional programming paradigm, *machine learning* is a family of techniques for inferring a program or algorithm from example data, rather than implementing it directly. So, with machine learning, you still feed your computer data, but instead of imposing instructions and expecting output, *you provide the expected output and let the machine find an algorithm by itself.*

To build a computer program that can identify who's in a photo, you can apply an algorithm that analyzes a large collection of images of your friend and generates a function that matches them. If you do this correctly, the generated function will also match new photos that you've never seen before. Of course, the program will have no knowledge of its purpose; all it can do is identify things that are similar to the original images you fed it.

In this situation, you call the images you provide the machine *training data*, and the names of the person on the picture *labels*. After you've *trained* an algorithm for your purpose, you can use it to *predict* labels on new data to test it. Figure 1.2 displays this example alongside a schema of the machine-learning paradigm.

Machine learning comes in when rules aren't clear; it can solve problems of the "I'll know it when I see it" variety. Instead of programming the function directly, you provide data that indicates what the function should do, and then methodically generate a function that matches your data.

In practice, you usually combine machine learning with traditional programming to build a useful application. For our face-detection app, you have to instruct the computer on how to find, load, and transform the example images before you can apply a machine-learning algorithm. Beyond that, you might use hand-rolled heuristics to separate headshots from photos of sunsets and latte art; then you can apply machine learning to put names to faces. Often a mixture of traditional programming techniques and advanced machine-learning algorithms will be superior to either one alone.

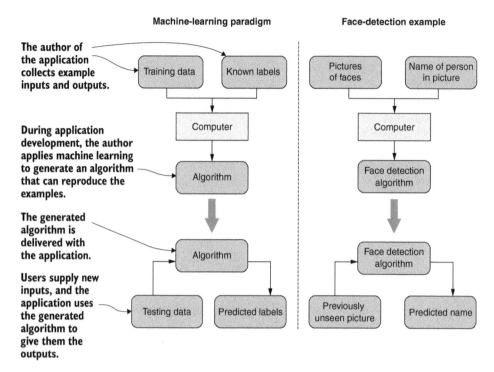

Figure 1.2   The machine-learning paradigm: during development, you generate an algorithm from a data set, and then incorporate that into your final application.

1.1.1   How does machine learning relate to AI?

Artificial intelligence, in the broadest sense, refers to any technique for making computers imitate human behavior. AI includes a huge range of techniques, including the following:

- Logic production systems, which apply formal logic to evaluate statements
- Expert systems, in which programmers try to directly encode human knowledge into software
- Fuzzy logic, which defines algorithms to help computers process imprecise statements

These sorts of rules-based techniques are sometimes called *classical AI* or *GOFAI* (good old-fashioned AI).

Machine learning is just one of many fields in AI, but today it's arguably the most successful one. In particular, the subfield of deep learning is behind some of the most exciting breakthroughs in AI, including tasks that eluded researchers for decades. In classical AI, researchers would study human behavior and try to encode rules that match it. Machine learning and deep learning flip the problem on its head: now you collect examples of human behavior and apply mathematical and statistical techniques to extract the rules.

Deep learning is so ubiquitous that some people in the community use *AI* and *deep learning* interchangeably. For clarity, we'll use *AI* to refer to the general problem of imitating human behavior with computers, and *machine learning* or *deep learning* to refer to mathematical techniques for extracting algorithms from examples.

1.1.2 *What you can and can't do with machine learning*

Machine learning is a specialized technique. You wouldn't use machine learning to update database records or render a user interface. Traditional programming should be preferred in the following situations:

- *Traditional algorithms solve the problem directly.* If you can directly write code to solve a problem, it'll be easier to understand, maintain, test, and debug.
- *You expect perfect accuracy.* All complex software contains bugs. But in traditional software engineering, you expect to methodically identify and fix bugs. That's not always possible with machine learning. You can improve machine-learning systems, but focusing too much on a specific error often makes the overall system worse.
- *Simple heuristics work well.* If you can implement a rule that's good enough with just a few lines of code, do so and be happy. A simple heuristic, implemented clearly, will be easy to understand and maintain. Functions that are implemented with machine learning are opaque and require a separate training process to update. (On the other hand, if you're maintaining a complicated sequence of heuristics, that's a good candidate to replace with machine learning.)

Often there's a fine line between problems that are feasible to solve with traditional programming and problems that are virtually impossible to solve, even with machine learning. Detecting faces in images versus tagging faces with names is just one example we've seen. Determining what language a text is written in versus translating that text into a given language is another such example.

We often resort to traditional programming in situations where machine learning might help—for instance, when the complexity of the problem is extremely high. When confronted with highly complex, information-dense scenarios, humans tend to settle for rules of thumb and narratives: think macroeconomics, stock-market predictions, or politics. Process managers and so-called experts can often vastly benefit from enhancing their intuition with insights gained from machine learning. Often, real-world data has more structure than anticipated, and we're just beginning to harvest the benefits of automation and augmentation in many of these areas.

1.2 *Machine learning by example*

The goal of machine learning is to construct a function that would be hard to implement directly. You do this by selecting a *model*, a large family of generic functions. Then you need a procedure for selecting a function from that family that matches your goal; this process is called *training* or *fitting* the model. You'll work through a simple example.

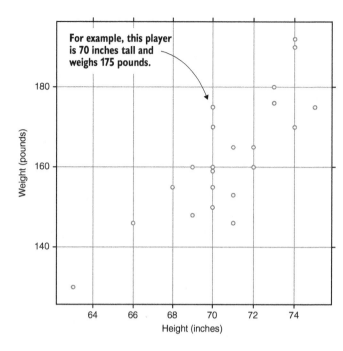

For example, this player
is 70 inches tall and
weighs 175 pounds.

Weight (pounds)

180

160

140

64   66   68   70   72   74

Height (inches)

**Figure 1.3   A simple example
data set. Each point on the
graph represents a soccer
player's height and weight.
Your goal is to fit a model to
these points.**

Let's say you collect the height and weight of some people and plot those values on a
graph. Figure 1.3 shows some data points that were pulled from the roster of a professional soccer team.

Suppose you want to describe these points with a mathematical function. First,
notice that the points, more or less, make a straight line going up and to the right. If
you think back to high school algebra, you may recall that functions of the form $f(x) =
ax + b$ describe straight lines. You might suspect that you could find values of a and b so
that $ax + b$ matches your data points fairly closely. The values of a and b are the *parameters*, or *weights*, that you need to figure out. This is your model. You can write Python
code that can generate any function in this family:

```
class GenericLinearFunction:
    def __init__(self, a, b):
        self.a = a
        self.b = b

    def evaluate(self, x):
        return self.a * x + self.b
```

How would you find out the right values of a and b? You can use rigorous algorithms
to do this, but for a quick and dirty solution, you could just draw a line through your
graph with a ruler and try to work out its formula. Figure 1.4 shows such a line that follows the general trend of the data set.

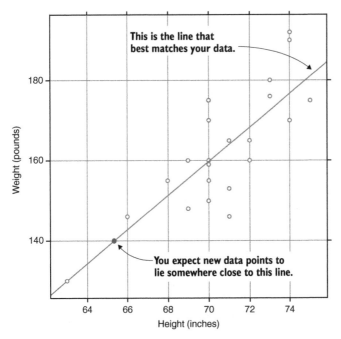

Figure 1.4 First you note that your data set roughly follows a linear trend, then you find the formula for a specific line that fits the data.

If you eyeball a couple of points that the line passes through, you can calculate a formula for the line; you'll get something like $f(x) = 4.2x - 137$. Now you have a specific function that matches your data. If you measure the height of a new person, you could then use your formula to estimate that person's weight. It won't be exactly right, but it may be close enough to be useful. You can turn your `GenericLinearFunction` into a specific function:

```
height_to_weight = GenericLinearFunction(a=4.2, b=-137)
height_of_new_person = 73
estimated_weight = height_to_weight.evaluate(height_of_new_person)
```

This should be a pretty good estimate, so long as your new person is also a professional soccer player. All the people in your data set are adult men, in a fairly narrow age range, who train for the same sport every day. If you try to apply your function to female soccer players, or Olympic weightlifters, or babies, you'll get wildly inaccurate results. Your function is only as good as your training data.

This is the basic process of machine learning. Here, your model is the family of all functions that look like $f(x) = ax + b$. And in fact, even something that simple is a useful model that statisticians use all the time. As you tackle more-complex problems, you'll use more-sophisticated models and more-advanced training techniques. But the core idea is the same: first describe a large family of possible functions and then identify the best function from that family.

Python and machine learning

All the code samples in this book are written in Python. Why Python? First, Python is an expressive high-level language for general application development. In addition, Python is among the most popular languages for machine learning and mathematical programming. This combination makes Python a natural choice for an application that integrates machine learning.

Python is popular for machine learning because of its amazing collection of numerical computing packages. Packages we use in this book include the following:

- *NumPy*—This library provides efficient data structures to represent numerical vectors and arrays, and an extensive library of fast mathematical operations. NumPy is the bedrock of Python's numerical computing ecosystem: every notable library for machine learning or statistics integrates with NumPy.
- *TensorFlow and Theano*—These are two graph computation libraries (*graph* in the sense of a network of connected steps, not *graph* as in *diagram*). They allow you to specify complex sequences of mathematical operations, and then generate highly optimized implementations.
- *Keras*—This is a high-level library for deep learning. It provides a convenient way for you to specify neural networks, and relies on TensorFlow or Theano to handle the raw computation.

We wrote the code examples in this book with Keras 2.2 and TensorFlow 1.8 in mind. You should be able to use any Keras version in the 2.*x* series with minimal modifications.

1.2.1 *Using machine learning in software applications*

In the previous section, you looked at a purely mathematical model. How can you apply machine learning to a real software application?

Suppose you're working on a photo-sharing app, in which users have uploaded millions of pictures with tags. You'd like to add a feature that suggests tags for a new photo. This feature is a perfect candidate for machine learning.

First, you have to be specific about the function you're trying to learn. Say you had a function like this:

```
def suggest_tags(image_data):
    """Recommend tags for an image.

    Input: image_data is a photo in bitmap format

    Returns: a ranked list of suggested tags
    """
```

Then the rest of the work is relatively straightforward. But it's not at all obvious how to start implementing a function like suggest_tags. That's where machine learning comes in.

If this were an ordinary Python function, you'd expect it to take some kind of Image object as input and perhaps return a list of strings as output. Machine-learning algorithms aren't so flexible about their inputs and outputs; they generally work on vectors and matrices. So as a first step, you need to represent your input and output mathematically.

If you resize the input photo to a fixed size—say, 128 × 128 pixels—then you can encode it as a matrix with 128 rows and 128 columns: one float value per pixel. What about the output? One option is to restrict the set of tags you'll identify; you could select perhaps the 1,000 most popular tags on the app. The output could then be a vector of size 1,000, where each element of the vector corresponds to a particular tag. If you allow the output values to vary anywhere between 0 and 1, you can generate ranked lists of suggested tags. Figure 1.5 illustrates this sort of mapping between concepts in your application and mathematical structures.

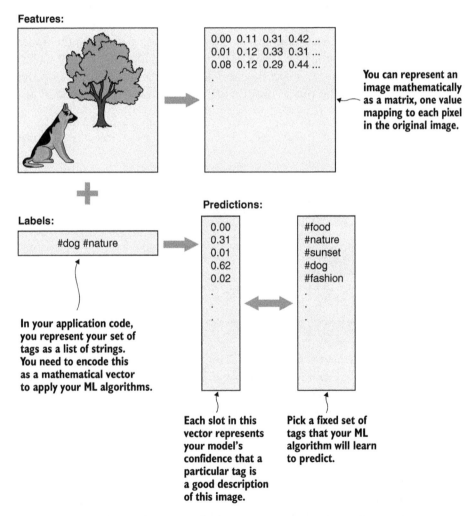

Figure 1.5 Machine-learning algorithms operate on mathematical structures, such as vectors and matrices. Your photo tags are stored in a standard computer data structure: a list of strings. This is one possible scheme for encoding that list as a mathematical vector.

This data preprocessing step you just carried out is an integral part of every machine-learning system. Usually, you load the data in raw format and carry out preprocessing steps to create *features*—input data that can be fed into a machine-learning algorithm.

1.2.2 Supervised learning

Next, you need an algorithm for training your model. In this case, you have millions of correct examples already—all the photos that users have already uploaded and manually tagged in your app. You can learn a function that attempts to match these examples as closely as possible, and you hope that it'll generalize to new photos in a sensible way. This technique is known as *supervised learning*, so-called because the *labels* of human-curated examples provide guidance for the training process.

When training is complete, you can deliver the final learned function with your application. Every time a user uploads a new photo, you pass it into the trained model function and get a vector back. You can match each value in the vector back to the tag it represents; then you can select the tags with the largest values and show them to the user. Schematically, the procedure you just outlined can be represented as shown in figure 1.6.

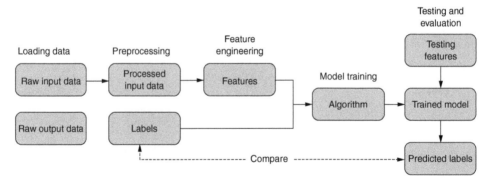

Figure 1.6   A machine-learning pipeline for supervised learning

How do you test your trained model? The standard practice is to set aside some of your original labeled data for that purpose. Before starting training, you can set aside a chunk of your data, say 10%, as a *validation set*. The validation set *isn't* included as part of the training data in any way. Then you can apply your trained model to the images in the validation set and compare the suggested tags to the known good tags. This lets you compute the accuracy of your model. If you want to experiment with different models, you have a consistent metric for measuring which is better.

In game AI, you can extract labeled training data from records of human games. And online gaming is a huge boon for machine learning: when people play a game online, the game server may save a computer-readable record. Examples of how to apply supervised learning to games are as follows:

- Given a collection of complete records of chess games, represent the game state in vector or matrix form and learn to predict the next move from data.
- Given a board position, learn to predict the likelihood of winning for that state.

1.2.3　*Unsupervised learning*

In contrast to supervised learning, the subfield of machine learning called *unsupervised learning* doesn't come with any labels to guide the learning process. In unsupervised learning, the algorithm has to learn to find patterns in the input data on its own. The only difference from figure 1.6 is that you're missing the labels, so you can't evaluate your predictions the way you did before. All other components stay the same.

An example of this is *outlier detection*—identifying data points that don't fit with the general trend of the data set. In the soccer player data set, outliers would indicate players who don't match the typical physique of their teammates. For instance, you could come up with an algorithm that measures the distance of a height-width pair to the line you eyeballed. If a data point exceeds a certain distance to the average line, you declare it an outlier.

In board-game AI, a natural question to ask is which pieces on the board belong together or form a group. In the next chapter, you'll see what this means for the game of Go in more detail. Finding groups of pieces that have a relationship is sometimes called *clustering* or *chunking*. Figure 1.7 shows an example of what this could look like for chess.

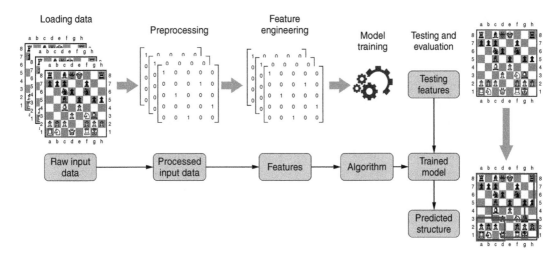

Figure 1.7　An unsupervised machine-learning pipeline for finding clusters or chunks of chess pieces

1.2.4　*Reinforcement learning*

Supervised learning is powerful, but finding quality training data can be a major obstacle. Suppose you're building a house-cleaning robot. The robot has various sensors that can detect when it's near obstacles, and motors that let it scoot around the floor and steer left or right. You need a control system: a function that can analyze the sensor input and decide how it should move. But supervised learning is impossible here. You have no examples to use as training data—your robot doesn't even exist yet.

Instead, you can apply *reinforcement learning*, a sort of trial-and-error approach. You start with an inefficient or inaccurate control system, and then you let the robot attempt its task. During the task, you record all the inputs your control system sees and the decisions it makes. When it's done, you need a way to evaluate how well it did, perhaps by calculating the fraction of the floor it vacuumed and how far it drained its battery. That whole experience gives you a small chunk of training data, and you can use it to improve the control system. By repeating the whole process over and over, you can gradually home in on an efficient control function. Figure 1.8 shows this process as a flowchart.

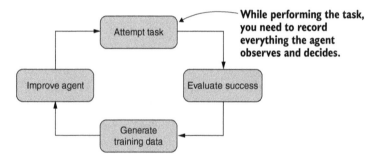

Figure 1.8   In reinforcement learning, agents learn to interact with their environment by trial and error. You repeatedly have your agent attempt its task to get a supervised signal to learn from. With every cycle, you can make an incremental improvement.

1.3   Deep learning

This book is made up of sentences. The sentences are made of words; the words are made of letters; the letters are made of lines and curves; and, ultimately, those lines and curves are made of tiny dots of ink. When teaching a child to read, you start with the smallest parts and work your way up: first letters, then words, then sentences, and finally complete books. (Normally, children learn to recognize lines and curves on their own.) This kind of hierarchy is the natural way for people to learn complex concepts. At each level, you ignore some detail, and the concepts become more abstract.

Deep learning applies the same idea to machine learning. Deep learning is a subfield of machine learning that uses a specific family of models: sequences of simple functions chained together. These chains of functions are known as *neural networks* because they were loosely inspired by the structure of natural brains. The core idea of deep learning is that these sequences of functions can analyze a complex concept as a hierarchy of simpler ones. The first layer of a deep model can learn to take raw data and organize it in basic ways—for example, grouping dots into lines. Each successive layer organizes the previous layer into more-advanced and more-abstract concepts. The process of learning these abstract concepts is called *representation learning*.

The amazing thing about deep learning is that you don't need to know what the intermediate concepts are in advance. If you select a model with enough layers and provide enough training data, the training process will gradually organize the raw data into increasingly high-level concepts. But how does the training algorithm know what concepts to use? It doesn't; it just organizes the input in any way that helps it to better match the training examples. There's no guarantee that this representation matches the way humans would think about the data. Figure 1.9 shows how representation learning fits into the supervised learning flow.

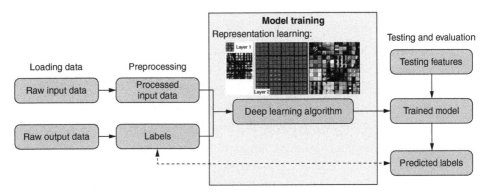

Figure 1.9   Deep learning and representation learning

All this power comes with a cost. Deep models have huge numbers of weights to learn. Recall the simple $ax + b$ model you used for your height and weight data set; that model had just two weights to learn. A deep model suitable for your image-tagging app could have a million weights. As a result, deep learning demands larger data sets, more computing power, and a more hands-on approach to training. Both techniques have their place. Deep learning is a good choice in the following circumstances:

- *Your data is in an unstructured form.* Images, audio, and written language are good candidates for deep learning. It's possible to apply simple models to that kind of data, but it generally requires sophisticated preprocessing.
- *You have large amounts of data available or have a plan for acquiring more.* In general, the more complex your model is, the more data you need to train it.
- *You have plenty of computing power or plenty of time.* Deep models involve more calculation for both training and evaluation.

You should prefer traditional models with fewer parameters in the following cases:

- *You have structured data.* If your inputs look more like database records, you can often apply simple models directly.
- *You want a descriptive model.* With simple models, you can look at the final learned function and examine how an individual input affects the output. This can give you insight about how the real-world system you're studying works. In deep models, the connection between a specific piece of the input and the final output is long and winding; it's difficult to interpret the model.

Because *deep learning* refers to the type of model you use, you can apply deep learning to any of the major machine-learning branches. For example, you can do supervised learning with a deep model or a simple model, depending on the type of training data you have.

1.4   What you'll learn in this book

This book provides a practical introduction to deep learning and reinforcement learning. To get the most out of this book, you should be comfortable reading and writing Python code, and have some familiarity with linear algebra and calculus. In this book, we teach the following:

- How to design, train, and test neural networks by using the Keras deep-learning library
- How to set up supervised deep-learning problems
- How to set up reinforcement-learning problems
- How to integrate deep learning with a useful application

Throughout the book, we use a concrete and fun example: building an AI that plays Go. Our Go bot combines deep learning with standard computer algorithms. We'll use straightforward Python to enforce the rules of the game, track the game state, and look ahead through possible game sequences. Deep learning will help the bot identify which moves are worth examining and evaluate who's ahead during a game. At each stage, you can play against your bot and watch it improve as you apply more-sophisticated techniques.

If you're interested in Go specifically, you can use the bot you'll build in the book as a starting point for experimenting with your own ideas. You can adapt the same techniques to other games. You'll also be able to add features powered by deep learning to other applications beyond games.

1.5   Summary

- Machine learning is a family of techniques for generating functions from data instead of writing them directly. You can use machine learning to solve problems that are too ambiguous to solve directly.
- Machine learning generally involves first choosing a *model*—a generic family of mathematical functions. Next you *train* the model—apply an algorithm to find the best function in that family. Much of the art of machine learning lies in selecting the right model and transforming your particular data set to work with it.
- Three of the major areas of machine learning are supervised learning, unsupervised learning, and reinforcement learning.
- Supervised learning involves learning a function from examples you already know to be correct. When you have examples of human behavior or knowledge available, you can apply supervised learning to imitate them on a computer.

- Unsupervised learning involves extracting structure from data without knowing what the structure is in advance. A common application is splitting a data set into logical groups.
- Reinforcement learning involves learning a function through trial and error. If you can write code to evaluate how well a program achieves a goal, you can apply reinforcement learning to incrementally improve a program over many trials.
- Deep learning is machine learning with a particular type of model that performs well on unstructured inputs, such as images or written text. It's one of the most exciting fields in computer science today; it's constantly expanding our ideas about what computers can do.

Go as a
machine-learning problem

2

This chapter covers

- Why are games a good subject for AI?
- Why is Go a good problem for deep learning?
- What are the rules of Go?
- What aspects of game playing can you solve with machine learning?

2.1 Why games?

Games are a favorite subject for AI research, and it's not just because they're fun. They also simplify some of the complexities of real life, so you can focus on the algorithms you're studying.

Imagine you see a comment on Twitter or Facebook: something like, "Ugh, I forgot my umbrella." You'd quickly conclude that your friend got caught out in the rain. But that information isn't included anywhere in the sentence. How did you reach that conclusion? First, you applied common knowledge about what umbrellas are for. Second, you applied social knowledge about the kinds of comments people bother to make: it'd be strange to say, "I forgot my umbrella" on a bright, sunny day.

As humans, we effortlessly factor in all this context when reading a sentence. This isn't so easy for computers. Modern deep-learning techniques are effective at processing the information you supply them. But you're limited in your ability to find all the relevant information and feed it to computers. Games sidestep that problem. They take place in an artificial universe, where all the information you need in order to make a decision is spelled out in the rules.

Games are especially well suited for reinforcement learning. Recall that reinforcement learning requires repeatedly running your program and evaluating how well it has accomplished a task. Imagine you're using reinforcement learning to train a robot to move around a building. Before the control system is finely tuned, you risk the robot falling down a flight of stairs or knocking over your furniture. Another option is to build a computer simulation of the environment in which the robot will operate. This eliminates the risks of letting an untrained robot run around in the real world but creates new problems. First, you have to invest in developing a detailed computer simulation, which is a significant project in its own right. Second, there's always a chance that your simulation isn't completely accurate.

With games, on the other hand, all you need to do is have your AI play. If it loses a few hundred thousand matches while it's learning, so what? In reinforcement learning, games are essential to serious research. Many cutting-edge algorithms were first demonstrated on Atari video games such as Breakout.

To be clear, you *can* successfully apply reinforcement learning to problems in the physical world. Many researchers and engineers have done so. But starting with games solves the problem of creating a realistic training environment and lets you focus on the mechanics and principles of reinforcement learning.

In this chapter, we introduce the rules of the game of Go. Next, we describe the structure of board-game AI at a high level, and identify points where you can introduce deep learning. Finally, we cover how you can evaluate the progress of your game AI throughout development.

2.2 A lightning introduction to the game of Go

You don't need to be a strong Go player to read this book, but you do need to understand the rules well enough to enforce them in a computer program. Fortunately, the rules are famously simple. In short, two players alternate placing black and white stones on a board, starting with the black player. The goal is to control as much of the board as possible with your own stones.

Although the rules are simple, Go strategy has endless depth, and we don't even attempt to cover it in this book. If you're interested in learning more, we provide some resources at the end of this section.

2.2.1 Understanding the board

A Go board is a square grid, shown in figure 2.1. Stones go on the intersections, not inside the squares. The standard board is 19 × 19, but sometimes players use a smaller board for a quick game. The most popular smaller options are 9 × 9 and 13 × 13 boards. (The size refers to the number of intersections on the board, not the number of squares.)

Notice that nine points are marked with a dot. These points are called the *star points*. Their main purpose is to help players judge distances on the board; they have no effect on game play.

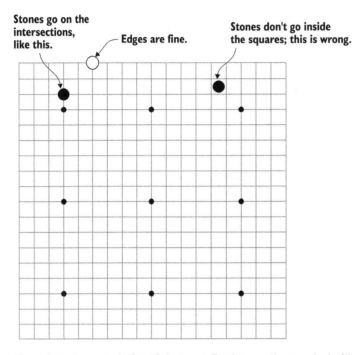

Stones go on the intersections, like this.   — **Edges are fine.**   **Stones don't go inside the squares; this is wrong.**

Figure 2.1   A standard 19 × 19 Go board. The intersections marked with the dots are the star points, which are solely for players' reference. Stones go on the intersections.

2.2.2   *Placing and capturing stones*

One player plays with black stones, and the other plays with white stones. The two players alternate placing stones on the board, starting with the black player. Stones don't move after they're on the board, although they can be captured and removed entirely. To capture your opponent's stones, you must completely surround them with your own. Here's how that works.

Stones of the same color that are touching are considered connected together, as shown in figure 2.2. For the purposes of connection, we consider only straight up, down, left, or right; diagonals don't count. Any empty point touching a connected group is called a *liberty* of that group. Every group needs at least

Figure 2.2   The three black stones are connected. They have four liberties on the points marked with squares. White can capture the black stones by placing white stones on all the liberties.

one liberty to stay on the board. You can capture your opponent's stones by filling their liberties.

When you place a stone in the last liberty of an opponent's group, that group is captured and removed from the board. The newly empty points are then available for either player to play on (so long as the move is legal). On the flip side, you may not play a stone that would have zero liberties, *unless you're completing a capture.*

An interesting consequence arises from the capturing rules. If a group of stones has two completely separate internal liberties, it can never be captured. See figure 2.3: black can't play at A, because that black stone would have no liberties and its placement wouldn't complete a capture because of the remaining liberty at B. Nor can black play at B, for the same reason. So black has no way to fill the last two liberties of the white

group. These internal liberties are called *eyes.* In contrast, black can play at C to capture five white stones, because even though that black stone would have no liberties, it completes the capture. That white group has only one eye and is doomed to get captured at some point.

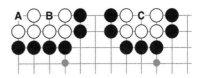

Figure 2.3 The white stones on the left can never be captured: black can play at neither A nor B. A black stone there would have no liberties, and is therefore an illegal play. On the other hand, black can play at C to capture five white stones.

Although it's not explicitly part of the rules, the idea that a group with two eyes can't be captured is the most basic part of Go strategy. In fact, this is the only strategy you'll specifically code into your bot's logic. All the more advanced Go strategies will be inferred through machine learning.

2.2.3 Ending the game and counting

Either player may pass any turn instead of placing a stone. When both players pass consecutive turns, the game is over. Before scoring, the players identify any dead stones: stones that have no chance of making two eyes or connecting up to friendly stones. Dead stones are treated exactly the same as captures when scoring the game. If a disagreement occurs, the players can resolve it by resuming play. But this is rare: if the status of any group is unclear, players usually try to resolve it before passing.

The goal of the game is to control a larger section of the board than your opponent. There are two ways to add up the score, but they nearly always give the same result.

The most common counting method is *territory scoring.* In this case, you get one point for every point on the board that's completely surrounded by your own stones, plus one point for every opponent's stone that you captured. The player with more points is the winner.

An alternative counting method is *area scoring.* With area scoring, you get one point for every point of territory, plus another point for every stone you have on the board. Except in rare cases, you'll get the same winner by either method: if neither player passes early, the difference in captures will equal the difference in stones on the board.

Territory scoring is more common in casual play, but it turns out that area scoring is slightly more convenient for computers. Throughout the book, our AI assumes it's playing under area scoring, unless otherwise noted.

In addition, the white player gets extra points as compensation for going second. This compensation is called *komi*. Komi is usually 6.5 points under territory scoring or 7.5 points under area scoring—the extra half point ensures there are no ties. We assume 7.5 points komi throughout the book.

Figure 2.4 shows the final position of an example 9 × 9 game.

Here's how you score this game:

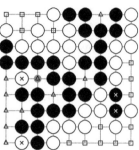

Figure 2.4   The final positions of a 9 × 9 game. Dead stones are marked with an ×. Black territory is marked with a triangle. White territory is marked with a square.

1 The stones marked with an X are considered dead: they count as captures, even though the players didn't make the capture in the game. Black also made a capture earlier in the game (not shown). So black has 3 captures, and white has 2 captures.

2 Black has 12 points of territory: the 10 points marked with a triangle, plus the 2 points underneath the dead white stones.

3 White has 17 points of territory: the 15 points marked with a square, plus the 2 points underneath the dead black stones.

4 Black has 27 stones on the board, after removing the dead black stones.

5 White has 25 stones on the board, after removing the dead white stones.

6 By territory scoring, white has 17 points of territory + 2 captures + 6.5 komi for a total of 25.5 points. Black has 12 points of territory + 3 captures for a total of 15 points.

7 By area scoring, white has 17 points of territory + 25 stones on the board + 7.5 komi for a total of 49.5 points. Black has 12 points of territory + 27 stones on the board for a total of 39 points.

8 By either counting method, white wins by 10.5 points.

A game can end in one more way: either player can resign at any point. In a game between experienced players, it's considered courteous to resign when you're clearly behind. For our AI to be a good opponent, it should learn to detect when it should resign.

2.2.4   *Understanding ko*

One more restriction exists on where you can place stones. To ensure that the game ends, it's illegal to play a move that will return the board to a previous state. Figure 2.5 shows an example of how this can happen.

In the diagram, black has just captured one stone. White might like to play at the point marked A to recapture the new black stone. But that would put the board back in the same position it was in two moves ago. Instead, white must play somewhere else on the board first. After that, if white captures at A, the global board position is different, so it's legal. Of course, this gives black the opportunity to protect the vulnerable stone. In order to recapture the black stone, the white player must create a distraction big enough to draw black's attention elsewhere on the board.

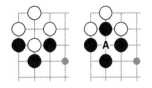

Figure 2.5   An illustration of the ko rule. First, black captures one white stone. White would like to capture back by playing at A, but that would revert the board to the previous position. The ko rule forbids such a play, in order to prevent an endless cycle of capture and recapture. White must play somewhere else on the board first. After that, the overall board position is new, so white can come back to capture at A later— assuming black doesn't protect it first.

This situation is called a *ko*, from the Japanese word for *eternity*. Players adopt special strategies when a ko is on the board, and this was a weak spot for previous generations of Go-playing programs. In chapter 7, we show how to give your neural networks a hint to help them learn ko tactics. This is a general technique for training neural networks effectively. Even when you can't articulate the rules that you want the neural network to learn, you can encode your inputs in a way that emphasizes the situations you want it to pay attention to.

2.3   *Handicaps*

When two players of unequal strength play, a simple system keeps the game interesting. The weaker player takes black and gets to place a few stones on the board before the game starts; these stones are called handicap stones. Then the stronger player takes white and makes the first move. In addition, in a handicap game, komi is usually reduced to just half a point. Normally, the purpose of komi is to negate the advantage a player gets from playing first; but the point of a handicap is to give black an extra advantage, so the two would work at cross purposes. The remaining half point of komi is just to break ties.

It's traditional to place the handicap stones on the star points, but some players allow the black player to choose where to put them.

2.4   *Where to learn more*

Although we've covered the rules of the game, we haven't even scratched the surface of what makes Go so engrossing and addictive. That's beyond the scope of this book, but we encourage you to play a few games and learn more on your own. Here are a few resources for further exploration:

- The best way to get into Go is to dive in and start playing, and it's easier than ever to find a casual game online. The popular Online Go Server (http:// online-go.com) enables you to play directly in your web browser. Even if you just learned the rules, its ranking system will help you find a competitive game. Other popular Go servers include the KGS Go Server (http://gokgs.com) and Tygem (www.tygembaduk.com).

- Sensei's Library (https://senseis.xmp.net) is a wiki-style reference, full of strategy, tips, history, and trivia.
- Janice Kim's *Learn to Play Go* series ranks among the best English-language Go books. We highly recommend volumes 1 and 2 for complete beginners: they'll quickly get you to the point where you can make sense of a game.

2.5   What can we teach a machine?

Whether you're programming a computer to play Go or tic-tac-toe, most board-game AIs share a similar overall structure. In this section, we provide a high-level overview of that structure, and identify specific problems the AI needs to solve. Depending on the game, the best solutions may involve game-specific logic, machine learning, or both.

2.5.1   Selecting moves in the opening

Early in the game, it's difficult to evaluate a particular move because of the huge number of variations in the rest of the game. Both chess and Go AIs often use an opening book: a database of opening sequences taken from expert human games. To build this, you need a collection of game records from strong players. You analyze the game records, looking for common positions. In any common position, if a strong consensus exists about the next move—say, one or two moves account for 80% of the followups—you add those moves to the opening book. The bot can then consult the book when playing a game. If any early game position appears in the opening book, the bot just looks up the expert move.

In chess and checkers, where pieces are removed from the board as the game progresses, AIs also contain similar endgame databases: when just a few pieces remain on the board, you can calculate all the variations in advance. This technique doesn't apply to Go, where the board fills up toward the end.

2.5.2   Searching game states

The core idea behind board-game AI is tree search. Think about how humans play a strategy game. First, we consider a possible move for our next turn. Then we need to think of how our opponent is likely to respond, then plan how we'd respond to that, and so on. We read out the variation as far as we can, and then judge whether the result is good. Then we backtrack a bit and look at a different variation to see if it's better.

This closely describes how the tree-search algorithms used in game AI work. Of course, humans can keep only a few variations in their heads at once, while computers can juggle millions with no trouble. Humans make up for their lack of raw computing power with intuition. Experienced chess and Go players are scary good at spotting the handful of moves worth considering.

Ultimately, raw computing power won out in chess. But a Go AI that could compete with top human players took an interesting twist: bringing human intuition to computers.

2.5.3 Reducing the number of moves to consider

In the context of game tree search, the number of possible moves on a given turn is the branching factor.

In chess, the average branching factor is about 30. At the start of the game, each player has 20 legal options for the first move; the number increases a bit as the board opens up. At that scale, it's realistic to read out every single possible move to four or five moves ahead, and a chess engine will read out the more promising lines much deeper than that.

By comparison, the branching factor in Go is enormous. On the first move of the game, there are 361 legal moves, and the number decreases slowly. The average branching factor is around 250 valid moves per turn. Looking ahead just four moves requires evaluating nearly 4 billion positions. It's crucial to narrow down the number of possibilities. Table 2.1 shows how the branching factor affects the number of possible game positions by comparing chess to Go.

Table 2.1 Approximate number of possible board states in games

	Branching factor 30 (chess)	Branching factor 250 (Go)
After two moves	900	62,500
After three moves	27,000	15 million
After four moves	810,000	4 billion
After five moves	24 million	1 trillion

In Go, rules-based approaches to move selection turn out to be mediocre at this task: it's extremely difficult to write out rules that reliably identify the most important area of the board. But deep learning is perfectly suited to the problem. You can apply supervised learning to train a computer to imitate a human Go player.

You start with a large collection of game records between strong human players; online gaming servers are a great resource here. Then you replay all the games on a computer, extracting each board position and the following move. That's your training set. With a suitably deep neural network, it's possible to predict the human move with better than 50% accuracy. You can build a bot that just plays the predicted human move, and it's already a credible opponent. But the real power comes when you combine these move predictions with tree search: the predicted moves give you a ranked list of branches to explore.

2.5.4 Evaluating game states

The branching factor limits how far an AI can look ahead. If you could read out a hypothetical sequence all the way to the end of the game, you'd know who'd win; it's easy to decide whether that's a good sequence. But that's not practical in any game more complex than tic-tac-toe: the number of possible variations is just too large. At some point, you have to stop and pick one of the incomplete sequences you've looked at. To do so, you take the final board position you've read out and assign it a numerical score. Of all the variations you analyzed, you select the move that leads to the best-scoring position. Figuring out how to compute that score is the tricky part: that's the problem of position evaluation.

In chess AIs, position evaluation is based on logic that makes sense to chess players. You can start with simple rules like this: if you capture my pawn, and I capture your rook, that's good for me. Top chess engines go far beyond that, factoring sophisticated rules about where on the board the pieces ended up and what other pieces are blocking their movement.

In Go, position evaluation may be even more difficult than move selection. The goal of the game is to cover more territory, but it's surprisingly difficult to count territory: the boundaries tend to remain vague until late in the game. Counting captures doesn't help much either; sometimes you can make it all the way to the end of a game with only a couple of captures. This is another area where human intuition reigns supreme.

Deep learning was a major breakthrough here as well. The kinds of neural networks that are suitable for move selection can also be trained to evaluate board positions. Instead of training a neural network to predict what the next move will be, you train it to predict who will win. You can design the network so it expresses its prediction as a probability; that gives you a numeric score to evaluate the board position.

2.6   *How to measure your Go AI's strength*

As you work on your Go AI, you'll naturally want to know how strong it is. Most Go players are familiar with the traditional Japanese ranking system, so you want to measure your bot's strength on that scale. The only way to calibrate its level is by playing against other opponents; you can use other AIs or human players for that purpose.

2.6.1   *Traditional Go ranks*

Go players generally use the traditional Japanese system, where players are given kyu (beginner) or dan (expert) ranks. The dan levels, in turn, are divided into amateur dan and professional dan. The strongest kyu rating is 1 kyu, and larger numbers are weaker. Dan ranks go in the opposite direction: 1 dan is one level stronger than 1 kyu, and larger dan ranks are stronger. For amateur players, the scale traditionally tops out at 7 dan. Amateur players can get a rating from their regional Go association, and online servers will also track a rating for their players. Table 2.2 shows how the ranks stack up.

Table 2.2 Traditional Go ranks

25 kyu	Complete beginners who have just learned the rules
20 kyu to 11 kyu	Beginners
10 kyu to 1 kyu	Intermediate players
1 dan and up	Strong amateur players
7 dan	Top amateur players, close to professional strength
Professional 1 dan to professional 9 dan	World's strongest players

Amateur ranks are based on the number of handicap stones required to make up the difference in strength between two players. For example, if Alice is 2 kyu and Bob is 5 kyu, Alice will usually give Bob three handicap stones so that they have an equal chance of winning.

Professional ranks work a little differently: they're more like titles. A regional Go association awards top players a professional rank based on results in major tournaments, and that rank is held for life. The amateur and professional scales aren't directly comparable, but you can safely assume that any player with a professional rank is at least as strong as an amateur 7 dan player. The top pros are significantly stronger than that.

2.6.2 *Benchmarking your Go AI*

An easy way to estimate the strength of your own bots is to play against other bots of known strength. Open source Go engines such as GNU Go and Pachi provide good benchmarks. GNU Go plays at around a 5 kyu level, and Pachi is about 1 dan (Pachi's level varies a bit depending on the amount of computing power you provide it). So if you have your bot play GNU Go 100 times, and it wins about 50 games, you can conclude that your bot is also somewhere in the neighborhood of 5 kyu.

To get a more precise rank, you can set up your AI to play on a public Go server with a rating system. A few dozen games should be enough to get a reasonable estimate.

2.7 *Summary*

- Games are a popular subject for AI research because they create controlled environments with known rules.
- The strongest Go AIs today rely on machine learning rather than game-specific knowledge. In part because of the huge number of possible variations to consider, rule-based Go AIs were historically not strong.
- Two places you can apply deep learning in Go are move selection and position evaluation.

- Move selection is the problem of narrowing the set of moves you need to consider in a particular board position. Without good move selection, your Go AI will have too many branches to read.
- Position evaluation is the problem of estimating which player is ahead and by how much. Without good position evaluation, your Go AI will have no ability to select a good variation.
- You can measure the strength of your AI by playing against widely available bots of known strength, such as GNU Go or Pachi.

Implementing
your first Go bot

3

This chapter covers

- Implementing a Go board by using Python
- Placing sequences of stones and simulating a game
- Encoding Go rules for this board to ensure legal moves are played
- Building a simple bot that can play against a copy of itself
- Playing a full game against your bot

In this chapter, you'll build a flexible library that provides data structures to represent Go games and algorithms that enforce the Go rules. As you saw in the preceding chapter, the rules of Go are simple, but in order to implement them on a computer, you have to consider all the edge cases carefully. If you're a novice to the game of Go or need a refresher of the rules, make sure you've read chapter 2. This chapter is technical and requires a good working knowledge of the Go rules to fully appreciate the details.

Representing the Go rules is immensely important, because it's the foundation for creating smart bots. Your bot needs to understand legal and illegal moves before you can teach it good and bad moves.

At the end of this chapter, you'll have implemented your first Go bot. This bot is still weak but has all the knowledge about the game of Go it needs to evolve into much stronger versions in the following chapters.

You'll start by formally introducing the board and fundamental concepts used to play a game of Go with a computer: what is a player, a stone, or a move? Next, you'll be concerned with game-play aspects. How can a computer quickly check which stones need to be captured or when the ko rule applies? When and how does a game come to an end? We'll answer all these questions throughout this chapter.

3.1   *Representing a game of Go in Python*

The game of Go is played on a square board. Usually, beginners start playing on a 9 × 9 or 13 × 13 board, and advanced and pro players play on a 19 × 19 board. But in principle, Go can be played on a board of any size. Implementing a square grid for the game is fairly simple, but you'll need to take care of a lot of intricacies down the line.

You represent a Go game in Python by building a module we'll call *dlgo* step-by-step. Throughout the chapter, you'll be asked to create files and implement classes and functions that will eventually lead to your first bot. All the code from this and later chapters can be found on GitHub at http://mng.bz/gYPe.

Although you should definitely clone this repository for reference, we strongly encourage you to follow along by creating the files from scratch to see how the library builds up piece by piece. The master branch of our GitHub repository contains all the code used in this book (and more). From this chapter on, there's additionally a specific Git branch for each chapter that has only the code you need for the given chapter. For instance, the code for this chapter can be found in branch chapter_3. The next chapters follow the same naming convention. Note that we've included extensive tests for most of the code found here and in later chapters in the GitHub repository.

To build a Python library to represent Go, you need a data model that's flexible enough to support a few use cases:

- Track the progress of a game you're playing against a human opponent.
- Track the progress of a game between two bots. This might seem to be exactly the same as the preceding point, but as it turns out, a few subtle differences exist. Most notably, naive bots have a hard time recognizing when the game is over. Playing two simple bots against each other is an important technique used in later chapters, so it's worth calling out here.
- Compare many prospective sequences from the same board position.
- Import game records and generate training data from them.

We start with a few simple concepts, such as what a player or move is. These concepts lay the foundations for tackling all of the preceding tasks in later chapters.

First, create a new folder, dlgo, and place an empty __init__.py file into it to initiate it as a Python module. Also, create two additional files called gotypes.py and goboard_slow.py in which you'll put all board- and game-play functionality. Your folder structure at this point should look as follows:

```
dlgo
    __init__.py
    gotypes.py
    goboard_slow.py
```

Black and white players take turns in Go, and you use enum to represent the different-colored stones. A Player is either black or white. After a player places a stone, you can switch the color by calling the other method on a Player instance. Put this Player class into gotypes.py.

Listing 3.1 Using an enum to represent players

```
import enum


class Player(enum.Enum):
    black = 1
    white = 2

    @property
    def other(self):
        return Player.black if self == Player.white else Player.white
```

As noted in the front matter, we're using Python 3 for this book. One of the reasons is that many modern aspects of the language, such as enums in gotypes.py, are part of the standard library in Python 3.

Next, to represent coordinates on the board, tuples are an obvious choice. The following Point class also goes into gotypes.py.

Listing 3.2 Using tuples to represent points of a Go board

```
from collections import namedtuple


class Point(namedtuple('Point', 'row col')):
    def neighbors(self):
        return [
            Point(self.row - 1, self.col),
            Point(self.row + 1, self.col),
            Point(self.row, self.col - 1),
            Point(self.row, self.col + 1),
        ]
```

A namedtuple lets you access the coordinates as point.row and point.col instead of point[0] and point[1], which makes for much better readability.

You also need a structure to represent the actions a player can take on a turn. Normally, a turn involves placing a stone on the board, but a player can also pass or resign at any time. Following American Go Association (AGA) conventions, we use the term *move* to mean any of those three actions, whereas a *play* refers to placing a stone. In the Move class, you therefore encode all three types of move (play, pass, resign) and make sure a move has precisely one of these types. For actual plays, you need to pass a Point to be placed. You add this Move class to the file goboard_slow.py.

Listing 3.3 Setting moves: plays, passes, or resigns

```
import copy
from dlgo.gotypes import Player         Any action a player can play
                                        on a turn—is_play, is_pass,
class Move():                           or is_resign—will be set.
    def __init__(self, point=None, is_pass=False, is_resign=False):
        assert (point is not None) ^ is_pass ^ is_resign
        self.point = point
        self.is_play = (self.point is not None)
        self.is_pass = is_pass
        self.is_resign = is_resign

    @classmethod                        This move places a
    def play(cls, point):               stone on the board.
        return Move(point=point)

    @classmethod
    def pass_turn(cls):         ◁──── This move passes.
        return Move(is_pass=True)

    @classmethod                        This move resigns
    def resign(cls):                    the current game.
        return Move(is_resign=True)
```

In what follows, clients generally won't call the Move constructor directly. Instead, you usually call Move.play, Move.pass_turn, or Move.resign to construct an instance of a move.

Note that so far, the Player, Point, and Move classes are all plain data types. Although they're fundamental to representing the board, they don't contain any game logic. This is done on purpose, and you'll benefit from separating game-play concerns like this.

Next, you implement classes that can update the game state by using the preceding three classes:

- The Board class is responsible for the logic of placing and capturing stones.
- The GameState class includes all the stones of the board, as well as tracking whose turn it is and what the previous state was.

3.1.1 *Implementing the Go board*

Before turning to GameState, let's first implement the Board class. Your first idea might be to create a 19 × 19 array tracking the state of each point in the board, and that's a good starting point. Now, think about the algorithm for checking when you need to remove stones from the board. Recall that the number of liberties of a single stone is defined by the number of empty points in its direct neighborhood. If all four neighboring points are occupied by an enemy stone, the stone has no liberties left and is captured. For larger groups of connected stones, the situation is more difficult to check. For instance, after placing a black stone, you have to check all neighboring white stones to see whether black captured any stones that you have to remove. Specifically, you have to check the following:

1 You see whether any neighbors have any liberties left.
2 You check whether any of the neighbors' neighbors have any liberties left.
3 You examine the neighbors' neighbors' neighbors, and so forth.

This procedure could require hundreds of steps to finish. Imagine a long chain snaking through the opponent's territory on a board with 200 moves played already. To speed this up, you can explicitly track all directly connected stones as a unit.

3.1.2 *Tracking connected groups of stones in Go: strings*

You saw in the preceding section that viewing stones in isolation can lead to an increased computational complexity. Instead, you'll keep track of groups of connected stones of the same color *and their liberties* at the same time. Doing so is much more efficient when implementing game logic.

You call a group of connected stones of the same color a *string of stones*, or simply a *string*, as shown in figure 3.1. You can build this structure efficiently with the Python set type as in the following GoString implementation, which you also put into goboard_slow.py.

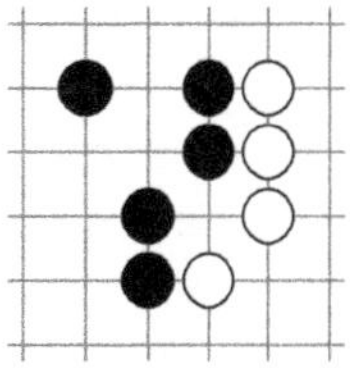

Figure 3.1 In this Go game, black has three strings of stones, and white has two. The large white string has six liberties, and the single white stone has only three.

Listing 3.4 Encoding strings of stones with set

```python
class GoString():
    def __init__(self, color, stones, liberties):
        self.color = color
        self.stones = set(stones)
        self.liberties = set(liberties)

    def remove_liberty(self, point):
        self.liberties.remove(point)

    def add_liberty(self, point):
```

Go strings are a chain of connected stones of the same color.

```
        self.liberties.add(point)

    def merged_with(self, go_string):                    Returns a new Go
        assert go_string.color == self.color             string containing all
        combined_stones = self.stones | go_string.stones  stones in both strings
        return GoString(
            self.color,
            combined_stones,
            (self.liberties | go_string.liberties) - combined_stones)

    @property
    def num_liberties(self):
        return len(self.liberties)

    def __eq__(self, other):
        return isinstance(other, GoString) and \
            self.color == other.color and \
            self.stones == other.stones and \
            self.liberties == other.liberties
```

Note that GoString directly tracks its own liberties, and you can access the number of liberties at any point by calling num_liberties, which is much more efficient than the preceding naive approach starting from individual stones.

Also, you have the ability to add and remove liberties from the given string by using remove_liberty and add_liberty. Liberties of a string will usually decrease when opponents play next to this string, and increase when this or another group captures opposing stones adjacent to this string.

Furthermore, note the merged_with method of GoString, which is called when a player connects two of its groups by placing a stone.

3.1.3 Placing and capturing stones on a Go board

After having discussed stones and strings of stones, the natural next step is to discuss how to place stones on the board. Using the GoString class from listing 3.4, the algorithm for placing a stone looks like this:

1 Merge any adjacent strings of the same color.
2 Reduce liberties of any adjacent strings of the opposite color.
3 If any opposite-color strings now have zero liberties, remove them.

Also, if the newly created string has zero liberties, you reject the move. This leads naturally to the following implementation of the Go Board class, which you also place at goboard_slow.py. You allow boards to have any number of rows or columns by instantiating them with num_rows and num_cols appropriately. To keep track of the board state internally, you use the private variable _grid, a dictionary you use to store strings of stones. First off, let's initiate a Go board instance by specifying its size.

Listing 3.5 Creating a Go Board instance

```
class Board():
    def __init__(self, num_rows, num_cols):
        self.num_rows = num_rows
        self.num_cols = num_cols
        self._grid = {}
```

> A board is initialized as an empty grid with the specified number of rows and columns.

Next, we discuss the Board method to place stones. In place_stone, you first have to inspect all neighboring stones of a given point for liberties.

Listing 3.6 Checking neighboring points for liberties

```
    def place_stone(self, player, point):
        assert self.is_on_grid(point)
        assert self._grid.get(point) is None
        adjacent_same_color = []
        adjacent_opposite_color = []
        liberties = []
        for neighbor in point.neighbors():
            if not self.is_on_grid(neighbor):
                continue
            neighbor_string = self._grid.get(neighbor)
            if neighbor_string is None:
                liberties.append(neighbor)
            elif neighbor_string.color == player:
                if neighbor_string not in adjacent_same_color:
                    adjacent_same_color.append(neighbor_string)
            else:
                if neighbor_string not in adjacent_opposite_color:
                    adjacent_opposite_color.append(neighbor_string)
        new_string = GoString(player, [point], liberties)
```

> First, you examine direct neighbors of this point.

Note that the first two lines in listing 3.6 use utility methods to check whether the point is within bounds for the given board and that the point hasn't been played yet. These two methods are defined as follows.

Listing 3.7 Utility methods for placing and removing stones

```
    def is_on_grid(self, point):
        return 1 <= point.row <= self.num_rows and \
            1 <= point.col <= self.num_cols

    def get(self, point):
        string = self._grid.get(point)
        if string is None:
            return None
        return string.color

    def get_go_string(self, point):
        string = self._grid.get(point)
        if string is None:
            return None
        return string
```

> Returns the content of a point on the board: a Player if a stone is on that point, or else None

> Returns the entire string of stones at a point: a GoString if a stone is on that point, or else None

Note that you also define `get_go_string` to return the string of stones associated with a given point. This functionality can be helpful in general, but it's particularly valuable to prevent *self-capture*, which we'll discuss in more detail in section 3.2.

Continuing with the definition of `place_stone` in listing 3.6, right after defining `new_string`, you follow the outlined three-step approach shown next.

Listing 3.8 Continuing our definition of `place_stone`

Merge any adjacent strings of the same color.
```
for same_color_string in adjacent_same_color:
    new_string = new_string.merged_with(same_color_string)
for new_string_point in new_string.stones:
    self._grid[new_string_point] = new_string
for other_color_string in adjacent_opposite_color:
    other_color_string.remove_liberty(point)
for other_color_string in adjacent_opposite_color:
    if other_color_string.num_liberties == 0:
        self._remove_string(other_color_string)
```
Reduce liberties of any adjacent strings of the opposite color.

If any opposite-color strings now have zero liberties, remove them.

Now, the only thing missing in our definition of a Go board is how to remove a string of stones, as required in `remove_string` in the last line of listing 3.8. This is fairly simple, as shown in listing 3.9, but you have to keep in mind that other stones might *gain* liberties when removing an enemy string. For instance, in figure 3.2, you can see that black can capture a white stone, thereby gaining one additional liberty for each black string of stones.

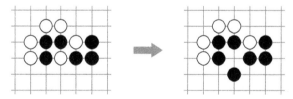

Black has two strings of stones: one with one liberty, and one with four liberties.

Black captures a stone, which adds a liberty to both strings.

Figure 3.2 Black can capture a white stone, thereby regaining a liberty for each of the Go strings adjacent to the captured stone.

Listing 3.9 Continuing our definition of `place_stone`

```
def _remove_string(self, string):
    for point in string.stones:
        for neighbor in point.neighbors():
            neighbor_string = self._grid.get(neighbor)
            if neighbor_string is None:
                continue
            if neighbor_string is not string:
                neighbor_string.add_liberty(point)
        self._grid[point] = None
```
Removing a string can create liberties for other strings.

This definition concludes our `Board` implementation.

3.2   *Capturing game state and checking for illegal moves*

Now that you've implemented the rules for placing and capturing stones on a Board, let's move on to playing games by capturing the current state of a game in a GameState class. Roughly speaking, *game state* knows about the board position, the next player, the previous game state, and the last move that has been played. What follows is just the beginning of the definition. You'll add more functionality to GameState throughout this section. Again, you put this into goboard_slow.py.

Listing 3.10   Encoding game state for a game of Go

```
class GameState():
    def __init__(self, board, next_player, previous, move):
        self.board = board
        self.next_player = next_player
        self.previous_state = previous
        self.last_move = move

    def apply_move(self, move):               ◄──── Returns the new
        if move.is_play:                            GameState after
            next_board = copy.deepcopy(self.board)  applying the move
            next_board.place_stone(self.next_player, move.point)
        else:
            next_board = self.board
        return GameState(next_board, self.next_player.other, self, move)

    @classmethod
    def new_game(cls, board_size):
        if isinstance(board_size, int):
            board_size = (board_size, board_size)
        board = Board(*board_size)
        return GameState(board, Player.black, None, None)
```

At this point, you can already decide when a game is over by adding the following code to your GameState class.

Listing 3.11   Deciding when a game of Go is over

```
    def is_over(self):
        if self.last_move is None:
            return False
        if self.last_move.is_resign:
            return True
        second_last_move = self.previous_state.last_move
        if second_last_move is None:
            return False
        return self.last_move.is_pass and second_last_move.is_pass
```

Now that you've implemented how to apply a move to the current game state, using apply_move, you should also write code to identify which moves are legal. Humans may accidentally attempt an illegal move. Bots might attempt illegal moves because they don't know any better. You need to check three rules:

- Confirm that the point you want to play is empty.
- Check that the move isn't a self-capture.
- Confirm that the move doesn't violate the ko rule.

Although the first point is trivial to implement, the other two deserve separate treatment, because they're rather tricky to handle properly.

3.2.1 Self-capture

When a string of your stones has only one liberty left and you play at the point that removes that liberty, we call that a *self-capture*. For instance, in figure 3.3, the black stones are doomed.

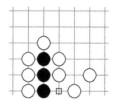

Figure 3.3 In this Go board state, the three black stones have one liberty left; namely, the marked point. You enforce the self-capture rule, so black isn't allowed to play there. White, on the other hand, can capture the three black stones by playing on the marked point.

White can capture them at any time by playing on the marked point, and black has no way to prevent it. But what if black played on the marked point? The entire group would have no liberties and would then be captured. Most rule sets forbid such a play, although a few exceptions exist. Most notably, self-capture is allowed in the quadrennial Ing Cup, which is one of the biggest prizes in international Go.

You'll enforce the self-capture rule in your code. This is consistent with the most popular rule sets, and it also reduces the number of moves your bots need to consider. It's possible to contrive a position where self-capture is the best move, but such situations are basically unheard of in serious games.

If you alter the surrounding stones in figure 3.3 slightly, you end up with a completely different situation, shown in figure 3.4.

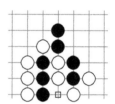

Figure 3.4 In this situation, the marked point is a capture for black, not suicide, because black will capture two white stones and thereby regain two liberties.

Note that in figure 3.4, and in general, you must remove opponent stones first before checking whether the newly played stone has any liberties. In all rule sets, this next move is a valid capture, not a self-capture, because black will regain two liberties by capturing the two white stones.

Note that the Board class does permit self-capture, but in GameState you'll enforce the rule by applying the move to a copy of the board and checking the number of liberties afterward.

Listing 3.12 Continuing our definition of `GameState` to enforce the self-capture rule

```
def is_move_self_capture(self, player, move):
    if not move.is_play:
        return False
    next_board = copy.deepcopy(self.board)
```

```
next_board.place_stone(player, move.point)
new_string = next_board.get_go_string(move.point)
return new_string.num_liberties == 0
```

3.2.2 Ko

Having checked for self-capture, you can now move on to implement the ko rule. Chapter 2 briefly covered ko and its importance in the game of Go. Roughly speaking, the ko rule applies if a move would return the board to the exact previous position. This doesn't imply that a player can't hit back immediately, as the following sequence of diagrams shows. In figure 3.5, white has just played the isolated stone on the bottom. Black's two stones now have only a single liberty left—but so does this white stone.

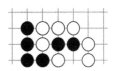

Figure 3.5 White wants to capture the two black stones in this situation, but white's stone has only one liberty left.

Black can now try to save its two stones by capturing this white stone, as shown in figure 3.6.

But white can immediately play at *the same point* that was played in figure 3.5, as shown in figure 3.7.

You can see that white can immediately recapture the three black stones, and the ko rule doesn't apply, because the overall board positions in figures 3.5 and 3.7 are different. This play is known as a *snapback*. In simple situations, the ko rule boils down to not being able to immediately recapture a stone. But snapbacks are common, and the existence of such positions shows that you have to be careful when implementing ko.

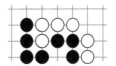

Figure 3.6 Continuing, black tries to rescue its two stones by capturing the isolated white stone.

You can specify the ko rule in many ways, but those ways are practically equivalent except in rare situations. The rule you'll enforce in your code is that a player may not play a stone that would re-create a previous game state, where the game state includes both the stones on the board and the player whose turn is next. This particular formulation is known as the *situational superko* rule.

Because each GameState instance keeps a pointer to the previous state, you can enforce the ko rule by walking back up the tree and checking the new state against the whole history. You do so by adding the following method to your GameState implementation.

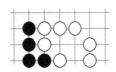

Figure 3.7 In this situation, white can just snap back (retake the three black stones) without violating the ko rule.

Listing 3.13 Does the current game state violate the ko rule?

```
@property
def situation(self):
    return (self.next_player, self.board)
```

```
def does_move_violate_ko(self, player, move):
    if not move.is_play:
        return False
    next_board = copy.deepcopy(self.board)
    next_board.place_stone(player, move.point)
    next_situation = (player.other, next_board)
    past_state = self.previous_state
    while past_state is not None:
        if past_state.situation == next_situation:
            return True
        past_state = past_state.previous_state
    return False
```

This implementation is simple and correct, but it's relatively slow. For each move, you create a deep copy of the board state and have to compare this state against all previous states, which adds up over time. In section 3.5, you'll encounter an interesting technique to speed up this step.

To wrap up your GameState definition, you can now decide whether a move is valid by using knowledge from section 3.2 about both ko and self-capture.

> **Listing 3.14   Is this move valid for the given game state?**

```
def is_valid_move(self, move):
    if self.is_over():
        return False
    if move.is_pass or move.is_resign:
        return True
    return (
        self.board.get(move.point) is None and
        not self.is_move_self_capture(self.next_player, move) and
        not self.does_move_violate_ko(self.next_player, move))
```

3.3   *Ending a game*

A key concept in computer Go is *self-play*. In self-play, you usually start with a weak Go-playing agent, have it play against itself, and use the game results to build a stronger bot. In chapter 4, you'll use self-play to evaluate board positions. In chapters 9 through 12, you'll use self-play to evaluate individual moves and the algorithms that selected them.

To take advantage of this technique, you need to make sure your self-play games end. Human games end when neither player can gain an advantage with their next move. This is a tricky concept even for humans. Beginners often end games by playing hopeless moves in the opponent's territory, or watching their opponent cut into what they believed to be solid territory. For computers, it's even more difficult. If our bot continues playing as long as legal moves remain available, it'll eventually fill up its own liberties and lose all of its stones.

You could think up some heuristics to help the bot finish the game in a sensible manner. For example:

- Don't play in a region that's completely surrounded by stones of the same color.
- Don't play a stone that would have only one liberty.
- Always capture an opposing stone with only one liberty.

Unfortunately, *all of those rules are too strict.* If our bots followed these rules, strong opponents would take advantage to kill groups that should live, rescue groups that should die, or simply gain a better position. Generally, our handcrafted rules should restrict the bot's options as little as possible, so that more sophisticated algorithms are free to learn the advanced tactics.

To solve this problem, you can look to the history of the game. In ancient times, the winner was simply the player with the most stones on the board. Players would end the game by filling every point they could, leaving only eyes for their groups empty. This could make the end of a game drag out for a long time, so players came up with a way to speed it up. If black clearly controlled a region of the board (any white stone played there would eventually get captured), there was no need for black to fill that region with stones. The players would just agree to count that area for black. This is where the concept of territory came from, and over centuries the rules evolved so that territory was counted explicitly.

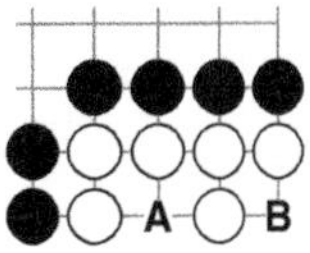

Figure 3.8 White has two eyes in the corner, at A and B, and shouldn't place a stone at either point. Otherwise, black can capture the whole group. Your naive bot won't be allowed to fill its own eye.

Scoring in this method avoids the question of what is territory and what isn't, but you still have to prevent your bot from killing its own stones; see figure 3.8.

You'll hardcode a rule that prevents the bot from filling in its own eyes, under the strictest possible definition. For our purposes, an *eye* is an empty point where all adjacent points and at least three out of four diagonally adjacent points are filled with friendly stones.

> **NOTE** Experienced Go players may notice that the preceding definition of *eye* will miss a valid eye in some cases. We'll accept those errors to keep the implementation simple.

You have to create a special case for eyes on the edge of the board; in that case, all the diagonally adjacent points must contain friendly stones.

You create a new submodule of dlgo called *agent* (by creating a new folder named *agent* and an empty __init__.py file within that folder) and place the following is_point_an_eye function into a file called helpers.py.

Listing 3.15 Is the given point on the board an eye?

```
from dlgo.gotypes import Point


def is_point_an_eye(board, point, color):
    if board.get(point) is not None:     ◁—— An eye is an empty point.
```

```
                   return False
All         ┌──▷ for neighbor in point.neighbors():
adjacent    │        if board.is_on_grid(neighbor):
points      │            neighbor_color = board.get(neighbor)
must        │            if neighbor_color != color:
contain     │                return False
friendly    │
stones.     └    friendly_corners = 0        ◁─────┤  We must control three out of four corners if
                 off_board_corners = 0              the point is in the middle of the board; on
                 corners = [                         the edge, you must control all corners.
                     Point(point.row - 1, point.col - 1),
                     Point(point.row - 1, point.col + 1),
                     Point(point.row + 1, point.col - 1),
                     Point(point.row + 1, point.col + 1),
                 ]
                 for corner in corners:
                     if board.is_on_grid(corner):
                         corner_color = board.get(corner)
                         if corner_color == color:
                             friendly_corners += 1
Point is    ┌        else:
on the      │            off_board_corners += 1
edge or     │    if off_board_corners > 0:
corner.     └──▷     return off_board_corners + friendly_corners == 4
                 return friendly_corners >= 3              ◁───── Point is in the middle.
```

You aren't explicitly concerned with determining the result of a game yet in this chapter, but counting points at the end of a game is definitely an important topic. Different tournaments and Go federations enforce slightly different rule sets. Throughout the book, you'll have your bots follow the AGA rules for *area counting*, also known as *Chinese counting*. Although *Japanese rules* are more popular in casual play, the AGA rules are a little easier for computers, and the rule differences rarely affect the outcome of a game.

3.4   Creating your first bot: the weakest Go AI imaginable

Having finished the implementation of the Go board and encoded game state, you can build your first Go-playing bot. This bot will be a weak player, but it'll lay the foundation for all of your improved bots to come. First, you define the interface that all of your bots will follow. You put the definition of an agent into base.py in the agent module.

Listing 3.16 Your central interface for Go agents

```
class Agent:
    def __init__(self):
        pass

    def select_move(self, game_state):
        raise NotImplementedError()
```

That's it, just this one method. All a bot does is select a move given the current game state. Of course, internally this may require other complex tasks such as evaluating the current position, but to play a game, that's all our bot will ever need.

Our first implementation will be as naive as possible: it'll randomly select any valid move that doesn't fill in one of its own eyes. If no such move exists, it'll pass. You place this random bot into naive.py under agents. Recall from chapter 2 that student ranks in Go usually range from 30 kyu to 1 kyu. By that scale, your random bot plays at the 30 kyu level, an absolute beginner.

Listing 3.17   A random Go bot, playing at about 30 kyu strength

```python
import random
from dlgo.agent.base import Agent
from dlgo.agent.helpers import is_point_an_eye
from dlgo.goboard_slow import Move
from dlgo.gotypes import Point


class RandomBot(Agent):
    def select_move(self, game_state):
        """Choose a random valid move that preserves our own eyes."""
        candidates = []
        for r in range(1, game_state.board.num_rows + 1):
            for c in range(1, game_state.board.num_cols + 1):
                candidate = Point(row=r, col=c)
                if game_state.is_valid_move(Move.play(candidate)) and \
                        not is_point_an_eye(game_state.board,
                                            candidate,
                                            game_state.next_player):
                    candidates.append(candidate)
        if not candidates:
            return Move.pass_turn()
        return Move.play(random.choice(candidates))
```

At this point, your module structure should look as follows (make sure to place an empty __init__.py in the folder to initialize the submodule):

```
dlgo
  ...
  agent
    __init__.py
    helpers.py
    base.py
    naive.py
```

Finally, you can set up a driver program that plays a full game between two instances of your random bot. First, you define convenient helper functions, such as printing the full board or an individual move on the console.

Go board coordinates can be specified in many ways, but in Europe it's most common to label the columns with letters of the alphabet, starting with A, and the rows with increasing numbers, starting at 1. In these coordinates, on a standard 19 × 19 board, the lower-left corner would be A1, and the top-right corner T19. Note that by convention the letter I is omitted to avoid confusion with 1.

You define a string variable COLS = 'ABCDEFGHJKLMNOPQRST', whose characters stand for the columns of the Go board. To display the board on the command line, you encode an empty field with a point (.), a black stone with an x, and a white one with an o. The following code goes into a new file we call utils.py in the dlgo package. You create a print_move function that prints the next move to the command line, and a print _board function that prints the current board with all its stones. You put this code in a file called bot_v_bot.py outside the dlgo module.

Listing 3.18   Utility functions for bot vs. bot games

```
from dlgo import gotypes

COLS = 'ABCDEFGHJKLMNOPQRST'
STONE_TO_CHAR = {
    None: ' . ',
    gotypes.Player.black: ' x ',
    gotypes.Player.white: ' o ',
}


def print_move(player, move):
    if move.is_pass:
        move_str = 'passes'
    elif move.is_resign:
        move_str = 'resigns'
    else:
        move_str = '%s%d' % (COLS[move.point.col - 1], move.point.row)
    print('%s %s' % (player, move_str))


def print_board(board):
    for row in range(board.num_rows, 0, -1):
        bump = " " if row <= 9 else ""
        line = []
        for col in range(1, board.num_cols + 1):
            stone = board.get(gotypes.Point(row=row, col=col))
            line.append(STONE_TO_CHAR[stone])
        print('%s%d %s' % (bump, row, ''.join(line)))
    print('    ' + '  '.join(COLS[:board.num_cols]))
```

You can set up a script that initiates two random bots that play each other on a 9 × 9 board until they decide the game is over.

Listing 3.19   A script to let a bot play against itself

```
from dlgo import agent
from dlgo import goboard
```

```
from dlgo import gotypes
from dlgo.utils import print_board, print_move
import time


def main():
    board_size = 9
    game = goboard.GameState.new_game(board_size)
    bots = {
        gotypes.Player.black: agent.naive.RandomBot(),
        gotypes.Player.white: agent.naive.RandomBot(),
    }
    while not game.is_over():
        time.sleep(0.3)
        print(chr(27) + "[2J")
        print_board(game.board)
        bot_move = bots[game.next_player].select_move(game)
        print_move(game.next_player, bot_move)
        game = game.apply_move(bot_move)

if __name__ == '__main__':
    main()
```

> **You set a sleep timer to 0.3 seconds so that bot moves aren't printed too fast to observe.**

> **Before each move, you clear the screen. This way, the board is always printed to the same position on the command line.**

You can start a bot game on the command line by running the following:

```
python bot_v_bot.py
```

You should see a lot of moves printed on the screen, and the game will end in both players passing. Recall that you encoded black stones as x, white stones as o, and empty points as a point (.). Here's an example of the last white move in a generated game:

```
9 o.oooooooo
8 ooooxxoxx
7 oooox.xxx
6 o.ooxxxxx
5 ooooxxxxx
4 ooooxxxxx
3 o.ooox.xx
2 ooooxxxxx
1 o.oooxxx.
  ABCDEFGHJ
Player.white passes
```

This bot is not only weak, but also a frustrating opponent: it'll keep stubbornly placing stones until the whole board is filled in, even if its position is hopeless. Moreover, no matter how often you let these bots play against each other, no *learning* is involved. This random bot will remain at its current level forever.

Throughout the rest of the book, you'll slowly improve on both of those weaknesses, building up an ever more interesting and powerful Go engine.

3.5   *Speeding up game play with Zobrist hashing*

Before wrapping up the chapter by describing how to play against your random bot, let's quickly address a speed issue in your current implementation by introducing an important technique. If you're not interested in ways to speed up your implementation, it's safe to skip to section 3.6 right away.

Recall from section 3.2 that to check for situational superko, you need to go through the entire history of positions of that game to see whether the current position has been there before. This is computationally expensive. To avoid this problem, you alter your setup slightly: instead of storing past board positions altogether, you simply store much smaller *hash values* of it.

Hashing techniques are omnipresent in computer science. One in particular is widely used in other games, such as chess: *Zobrist hashing* (named after computer scientist Albert Zobrist, who built one of the first Go bots in the early 1970s). In Zobrist hashing, you assign a hash value to each possible Go move on the board. For best results, you should choose each hash value randomly. In Go, each move is either black or white, so on a 19 × 19 board, a full Zobrist hash table consists of 2 × 19 × 19 = 722 hash values. You use this small number of 722 hashes representing individual moves to encode the most complex of board positions. Figure 3.9 shows how it works.

What's interesting about the procedure shown in figure 3.9 is that a full board state can be encoded by a single hash value. You start with the hash value of the empty board, which you can choose to be 0 for simplicity. The first move has a hash value, and you can apply this move to the board by carrying out an XOR operation of the board with the move hash. We call this operation *applying the hash*. Following this convention, for each new move, you can apply its current hash to the board. This allows you to track the current board state as one hash value.

Note that you can reverse any move by applying its hash again (a convenient feature of the XOR operation). We call this *unapplying the hash value*. This is important, because with this property, you can easily remove stones from the board when they're captured. For instance, if a single black stone at C3 on the board is captured, you can remove it from the current board state hash by applying the hash value of C3. Of course, in this scenario, the hash value of the white stone capturing C3 has to be applied as well. If a white move captures multiple black stones, you unapply all of their hashes.

Given that you've chosen your hash values sufficiently large and general, so that no hash collisions occur (two different game states never result in the same hash value), you can encode any board position like this. In practice, you don't check for hash collisions, but simply assume there are none.

To implement Zobrist hashing for your Go board implementation, you first need to generate the hashes. You generate 64-bit random integers by using Python's `random`

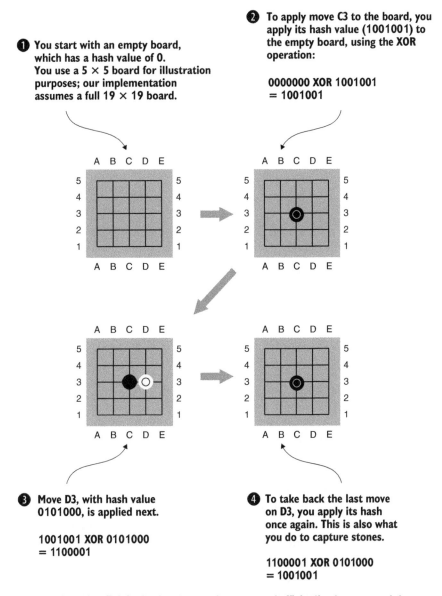

1 You start with an empty board, which has a hash value of 0. You use a 5 × 5 board for illustration purposes; our implementation assumes a full 19 × 19 board.

2 To apply move C3 to the board, you apply its hash value (1001001) to the empty board, using the XOR operation:

0000000 XOR 1001001
= 1001001

3 Move D3, with hash value 0101000, is applied next.

1001001 XOR 0101000
= 1100001

4 To take back the last move on D3, you apply its hash once again. This is also what you do to capture stones.

1100001 XOR 0101000
= 1001001

Figure 3.9 Using Zobrist hashes to encode moves and efficiently store game state

library for each of the $3 \times 19 \times 19$ possible point states. Note that in Python the symbol ^ carries out an XOR operation. For the empty board, you choose a value of 0.

```
import random

from dlgo.gotypes import Player, Point
```

```
def to_python(player_state):
    if player_state is None:
        return 'None'
    if player_state == Player.black:
        return Player.black
    return Player.white


MAX63 = 0x7fffffffffffffff

table = {}
empty_board = 0
for row in range(1, 20):
    for col in range(1, 20):
        for state in (Player.black, Player.white):
            code = random.randint(0, MAX63)
            table[Point(row, col), state] = code
print('from .gotypes import Player, Point')
print('')
print("__all__ = ['HASH_CODE', 'EMPTY_BOARD']")
print('')
print('HASH_CODE = {')
for (pt, state), hash_code in table.items():
    print('    (%r, %s): %r,' % (pt, to_python(state), hash_code))
print('}')
print('')
print('EMPTY_BOARD = %d' % (empty_board,))
```

Running this script prints the desired hashes on the command line. Executing the preceding code generates Python code printed to the command line. Place this output in the file zobrist.py in the dlgo module.

Now that you have the hashes available, all you need to do is to replace your old state-tracking mechanism by storing hashes instead. Create a copy of goboard_slow.py called goboard.py, in which you'll make all the necessary changes for the rest of this section. Alternatively, you can follow the code in goboard.py from our GitHub repository. You start with a slight modification to make GoString and both stones and liberties immutable, meaning they can't by modified after they're created. You can do this by using Python's frozenset instead of set. A frozenset doesn't have methods to add or remove items, so you need to create a new set instead of modifying an existing one.

Listing 3.21 GoString instances with immutable sets of stones and liberties

```
class GoString:
    def __init__(self, color, stones, liberties):
        self.color = color
        self.stones = frozenset(stones)
        self.liberties = frozenset(liberties)    ◁──

    def without_liberty(self, point):    ◁──
```

Stones and liberties are now immutable frozenset instances.

The without_liberty method replaces the previous remove_liberty method...

```
        new_liberties = self.liberties - set([point])
        return GoString(self.color, self.stones, new_liberties)

    def with_liberty(self, point):
        new_liberties = self.liberties | set([point])
        return GoString(self.color, self.stones, new_liberties)
```

... and with_liberty replaces add_liberty.

In `GoString`, you replace two methods to account for immutable state and leave the other helper methods, such as `merged_with` or `num_liberties`, untouched.

Next, you update relevant parts of the Go board. Remember to place all the code for the rest of this section in goboard.py, your copy of goboard_slow.py.

Listing 3.22 Instantiating the Go board with a _hash value for the empty board

```
from dlgo import zobrist

class Board:
    def __init__(self, num_rows, num_cols):
        self.num_rows = num_rows
        self.num_cols = num_cols
        self._grid = {}
        self._hash = zobrist.EMPTY_BOARD
```

Next, in your `place_stone` method, whenever a new stone is placed, you apply the hash of the respective color. Make sure to apply these changes to the goboard.py file as well, just as with every other piece of code in this section.

Listing 3.23 Placing a stone means applying the hash of that stone

You merge any adjacent strings of the same color.

Until this line, place_stone remains the same.

```
new_string = GoString(player, [point], liberties)

for same_color_string in adjacent_same_color:
    new_string = new_string.merged_with(same_color_string)
for new_string_point in new_string.stones:
    self._grid[new_string_point] = new_string
```

Next, you apply the hash code for this point and player.

```
self._hash ^= zobrist.HASH_CODE[point, player]
```

Then you reduce liberties of any adjacent strings of the opposite color.

```
for other_color_string in adjacent_opposite_color:
    replacement = other_color_string.without_liberty(point)
    if replacement.num_liberties:

        self._replace_string(other_color_string.without_liberty(point))
        else:
            self._remove_string(other_color_string)
```

If any opposite-color strings now have zero liberties, remove them.

To remove a stone, you apply its hash to the board once again.

Listing 3.24   Removing a stone means unapplying the hash value of the stone

```
def _replace_string(self, new_string):        ◁──┐  This new helper method
    for point in new_string.stones:               │  updates your Go board grid.
        self._grid[point] = new_string

def _remove_string(self, string):                    ┐  Removing a string can create
    for point in string.stones:                   ◁──┘  liberties for other strings.
        for neighbor in point.neighbors():
            neighbor_string = self._grid.get(neighbor)
            if neighbor_string is None:
                continue
            if neighbor_string is not string:
                self._replace_string(neighbor_string.with_liberty(point))
        self._grid[point] = None

        self._hash ^= zobrist.HASH_CODE[point, string.color]
```

With Zobrist hashing, you need to unapply the hash for this move.

The last thing you add to the `Board` class is a utility method to return the current Zobrist hash.

Listing 3.25   Returning the current Zobrist hash of the board

```
def zobrist_hash(self):
    return self._hash
```

Now that you've encoded your Go board with Zobrist hashing values, let's see how to improve `GameState` with it.

Before, the previous game state was set like this: `self.previous_state = previous`, which we argued was too expensive, because you had to cycle through all past states to check for ko. Instead, you want to store Zobrist hashes, and you do this by using the new variable `previous_states`, as shown in the next code listing.

Listing 3.26   Initializing game state with Zobrist hashes

```
class GameState:
    def __init__(self, board, next_player, previous, move):
        self.board = board
        self.next_player = next_player
        self.previous_state = previous
        if self.previous_state is None:
            self.previous_states = frozenset()
        else:
            self.previous_states = frozenset(
                previous.previous_states |
                {(previous.next_player, previous.board.zobrist_hash())})
        self.last_move = move
```

If the board is empty, `self.previous_states` is an empty immutable `frozenset`. Otherwise, you augment the states by a pair: the color of the next player and the Zobrist hash of the previous game state.

Having set up all this, you can finally drastically improve on your `does_move_violate_ko` implementation.

> **Listing 3.27 Fast checking of game states for ko with Zobrist hashes**

```
def does_move_violate_ko(self, player, move):
    if not move.is_play:
        return False
    next_board = copy.deepcopy(self.board)
    next_board.place_stone(player, move.point)
    next_situation = (player.other, next_board.zobrist_hash())
    return next_situation in self.previous_states
```

Checking previous board states by the single line `next_situation in self.previous _states` is an order of magnitude faster than the explicit loop over board states you had before.

This interesting hashing trick will enable much faster self-play in later chapters, leading to much quicker improvements in game play.

Further speeding up your Go board implementation

We gave an in-depth treatment of the original goboard_slow.py implementation and showed how to speed it up with Zobrist hashing to arrive at goboard.py. In the GitHub repository, you'll see yet another Go board implementation called goboard_fast.py, in which you can further speed up game play. These speed improvements are extremely valuable in later chapters, but come as a trade-off for readability.

If you're interested in how to make your Go board even quicker, look at goboard_fast .py and the comments found there. Most of the optimizations are tricks to avoid constructing and copying Python objects.

3.6   *Playing against your bot*

Having created a weak bot that plays against itself, you may wonder whether you can test your knowledge from chapter 2 to play against it yourself. This is indeed possible and doesn't require a lot of changes compared to the setup for bot versus bot.

You need one more utility function that you put into utils.py to help you read coordinates from human input.

> **Listing 3.28 Transforming human input into coordinates for your Go board**

```
def point_from_coords(coords):
    col = COLS.index(coords[0]) + 1
    row = int(coords[1:])
    return gotypes.Point(row=row, col=col)
```

This function transforms inputs like C3 or E7 into Go board coordinates. With this in mind, you can now set up your program for a 9 × 9 game in human_v_bot.py like this.

> **Listing 3.29   Setting up a script so you can play your own bot**

```
from dlgo import agent
from dlgo import goboard_slow as goboard
from dlgo import gotypes
from dlgo.utils import print_board, print_move, point_from_coords
from six.moves import input


def main():
    board_size = 9
    game = goboard.GameState.new_game(board_size)
    bot = agent.RandomBot()

    while not game.is_over():
        print(chr(27) + "[2J")
        print_board(game.board)
        if game.next_player == gotypes.Player.black:
            human_move = input('-- ')
            point = point_from_coords(human_move.strip())
            move = goboard.Move.play(point)
        else:
            move = bot.select_move(game)
        print_move(game.next_player, move)
        game = game.apply_move(move)


if __name__ == '__main__':
    main()
```

You, the human player, will play as black. Your random bot takes white. Start the script with the following:

```
python human_v_bot.py
```

You'll be prompted to type in a move and confirm with enter. For instance, if you choose to play G3 as your first move, the bot's response may look as follows:

```
Player.white D8
9 .........
8 ...o.....
7 .........
6 .........
5 .........
4 .........
3 ......x..
2 .........
1 .........
  ABCDEFGHJ
```

If you wish, you can continue and play a full game against this bot. But because the bot plays random moves, it isn't interesting to do so yet.

Note that in terms of following the Go rules, this bot is complete already. It knows everything about the game of Go it'll ever need to know. This fact is important, because it frees you to completely focus on the algorithms that improve game play from now on. This bot represents the baseline you start from. In the following chapters, you'll introduce more-interesting techniques to create stronger bots.

3.7 Summary

- The two players of a game of Go are best encoded by using an enum.
- A point on the Go board is characterized by its immediate neighborhood.
- A move in Go is either playing, passing, or resigning.
- Strings of stones are connected groups of stones of the same color. Strings are important to efficiently check for captured stones after placing a stone.
- The Go Board has all the logic of placing and capturing stones.
- In contrast, GameState keeps track of whose turn it is, the stones currently on the board, and the history of previous states.
- Ko can be implemented by using the situational superko rule.
- Zobrist hashing is an important technique to efficiently encode game-play history and speed up checking for ko.
- A Go-playing agent can be defined with one method: select_move.
- Your random bot can play against itself, other bots, and human opponents.

Part 2

Machine learning and game AI

In part 2, you'll learn the components of both classical and modern game AI. You'll start with a variety of tree search algorithms, which are essential tools in game AI and all kinds of optimization problems. Next, you'll learn about deep learning and neural networks, starting from the mathematical basics and working up to many practical design considerations. Finally, you'll get an introduction to reinforcement learning, a framework where your game AI can improve through practice.

Of course, none of these techniques are limited to games. Once you've mastered the components, you'll start seeing opportunities to apply them in all sorts of domains.

Playing games with tree search

4

This chapter covers

- Finding the best move with the minimax algorithm
- Pruning minimax tree search to speed it up
- Applying Monte Carlo tree search to games

Suppose you're given two tasks. The first is to write a computer program that plays chess. The second is to write a program that plans how to efficiently pick orders in a warehouse. What could these programs have in common? At first glance, not much. But if you step back and think in abstract terms, you can see a few parallels:

- *You have a sequence of decisions to make.* In chess, your decisions are about which piece to move. In the warehouse, your decisions are about which item to pick up next.
- *Early decisions can affect future decisions.* In chess, moving a pawn early on may expose your queen to a counterattack many turns later. In the warehouse, if you go for a widget on shelf 17 first, you may need to backtrack all the way to shelf 99 later.

- *At the end of a sequence, you can evaluate how well you achieved your goal.* In chess, when you reach the end of the game, you know who won. In the warehouse, you can add up the time it took to gather all the items.
- *The number of possible sequences can get enormous.* There are around 10^{100} ways a chess game can unfold. In the warehouse, if you have 20 items to pick up, there are 2 billion possible paths to choose from.

Of course, this analogy only goes so far. In chess, for example, you alternate turns with an opponent who is actively trying to thwart your intentions. That doesn't happen in any warehouse you'd want to work in.

In computer science, *tree-search* algorithms are strategies for looping over many possible sequences of decisions to find the one that leads to the best outcome. In this chapter, we cover tree-search algorithms as they apply to games; many of the principles can extend to other optimization problems as well. We start with the *minimax* search algorithm, in which you switch perspectives between two opposing players on each turn. The minimax algorithm can find perfect lines of play but is too slow to apply to sophisticated games. Next, we take a look at two techniques for getting a useful result while searching only a tiny fraction of the tree. One of these is *pruning*: you speed up the search by eliminating sections of the tree. To prune effectively, you need to bring real-world knowledge of the problem into your code. When that's not possible, you can sometimes apply *Monte Carlo tree search* (MCTS). MCTS is a randomized search algorithm that can find a good result without any domain-specific code.

With these techniques in your toolkit, you can start building AIs that can play a variety of board and card games.

4.1 Classifying games

Tree-search algorithms are mainly relevant to games where you take turns, and a discrete set of options is available on each turn. Many board and card games fit this description. On the other hand, tree search won't help a computer play basketball, charades, or World of Warcraft. We can further classify board and card games according to two useful properties:

- *Deterministic vs. nondeterministic*—In a *deterministic* game, the course of the game depends only on the players' decisions. In a *nondeterministic* game, an element of randomness is involved, such as rolling dice or shuffling cards.
- *Perfect information vs. hidden information*—In *perfect information* games, both players can see the full game state at all times; the whole board is visible, or everyone's cards are on the table. In *hidden information* games, each player can see only part of the game state. Hidden information is common in card games, where each player is dealt a few cards and can't see what the other players are holding. Part of the appeal of hidden information games is guessing what the other players know based on their game decisions.

Table 4.1 shows how several well-known games fit into this taxonomy.

Table 4.1   Classifying board and card games

	Deterministic	**Nondeterministic**
Perfect information	Go, chess	Backgammon
Hidden information	Battleship, Stratego	Poker, Scrabble

In this chapter, we primarily focus on deterministic, perfect information games. On each turn of such games, one move is theoretically the best. There's no luck and no secrets; before you choose a move, you know every move your opponent might choose in response, and everything you might do after that, and so on to the end of the game. In theory, you should have the whole game planned out on the first move. The minimax algorithm does exactly that in order to come up with the perfect play.

In reality, the games that have stood the test of time, such as chess and Go, have an enormous number of possibilities. To humans, every game seems to take on a life of its own, and even computers can't calculate them all the way to the end.

All the examples in this chapter contain little game-specific logic, so you can adapt them to any deterministic, perfect information game. To do so, you can follow the pattern of our goboard module and implement your new game logic in classes such as `Player`, `Move`, and `GameState`. The essential functions for `GameState` are `apply_move`, `legal_moves`, `is_over`, and `winner`. We have done this for tic-tac-toe; you can find this in the ttt module on GitHub (http://mng.bz/gYPe).

> **Games for AI experiments**
> Need some inspiration? Look up the rules for any of the following games:
>
> - Chess
> - Checkers
> - Reversi
> - Hex
> - Chinese checkers
> - Mancala
> - Nine Men's Morris
> - Gomoku

4.2   Anticipating your opponent with minimax search

How can you program a computer to decide what move to make next in a game? To start, you can think about how humans would make the same decision. Let's start with the simplest deterministic, perfect information game there is: tic-tac-toe. The technical name for the strategy we'll describe is *minimaxing*. This term is a contraction of *minimizing and maximizing*: you're trying to maximize your score, while your opponent is trying to minimize your score. You can sum up the algorithm in one sentence: assume that your opponent is as smart as you are.

Let's see how minimaxing works in practice. Take a look at figure 4.1. What move should X make next? There's no trick here; taking the lower-right corner wins the game. You can make that into a general rule: take any move that immediately wins the game. There's no way this plan can go wrong. You could implement this rule in code with something like the following listing.

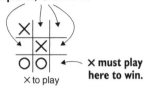

If X plays any of these squares, O can win.

X must play here to win.

X to play

Figure 4.1 What move should X make next? This is an easy one: playing in the lower-right corner wins the game.

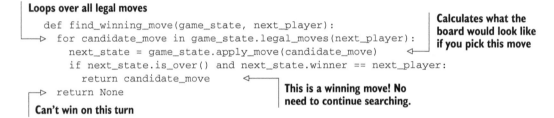

Listing 4.1 A function that finds a move that immediately wins the game

Loops over all legal moves

```
def find_winning_move(game_state, next_player):
    for candidate_move in game_state.legal_moves(next_player):
        next_state = game_state.apply_move(candidate_move)
        if next_state.is_over() and next_state.winner == next_player:
            return candidate_move
    return None
```

Calculates what the board would look like if you pick this move

This is a winning move! No need to continue searching.

Can't win on this turn

Figure 4.2 illustrates the hypothetical board positions this function would examine. This structure, in which a board position points to possible follow-ups, is called a *game tree*.

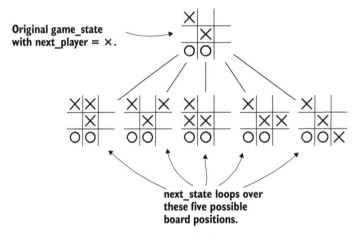

Original game_state with next_player = X.

next_state loops over these five possible board positions.

Figure 4.2 An illustration of an algorithm to find the winning move. You start with the position at the top. You loop over every possible move and calculate the game state that would result if you played that move. Then you check whether that hypothetical game state is a winning position for X.

Let's back up a bit. How did you get in this position? Perhaps the previous position looked like figure 4.3. The O player naively hoped to make three in a row across the bottom. But that assumes that X will cooperate with the plan. This gives a corollary to our previous rule: don't choose any move that gives your opponent a winning move. Listing 4.2 implements this logic.

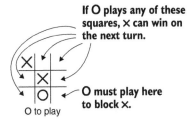

If O plays any of these squares, ✕ can win on the next turn.

O must play here to block ✕.

O to play

Figure 4.3 What move should O make next? If O plays in the lower left, you must assume that ✕ will follow up in the lower right to win the game. O must find the only move that prevents this.

Listing 4.2 A function that avoids giving the opponent a winning move

possible_moves will become a list of all moves worth considering.

```
def eliminate_losing_moves(game_state, next_player):
    opponent = next_player.other()
    possible_moves = []
    for candidate_move in game_state.legal_moves(next_player):
        next_state = game_state.apply_move(candidate_move)
        opponent_winning_move = find_winning_move(next_state, opponent)
        if opponent_winning_move is None:
            possible_moves.append(candidate_move)
    return possible_moves
```

Loops over all legal moves

Calculates what the board would look like if you play this move

Does this give your opponent a winning move? If not, this move is plausible.

Now, you know that you must block your opponent from getting into a winning position. Therefore, you should assume that your opponent is going to do the same to you. With that in mind, how can you play to win? Take a look at the board in figure 4.4.

If you play in the center, you have two ways to complete three in a row: top middle or lower right. The opponent can't block them both. We can describe this general principle as follows: look for a move that sets up a subsequent winning move that your opponent can't block. Sounds complicated, but it's easy to build this logic on top of the functions you've already written.

If ✕ plays here, that creates two ways to win.

✕ to play

Figure 4.4 What move should ✕ make? If ✕ plays in the center, there will be two ways to complete three in a row: (1) top middle and (2) lower right. O can block only one of these options, so ✕ is guaranteed a win.

Listing 4.3   A function that finds a two-move sequence that guarantees a win

Loops over all legal moves

```
def find_two_step_win(game_state, next_player):
    opponent = next_player.other()
    for candidate_move in game_state.legal_moves(next_player):
        next_state = game_state.apply_move(candidate_move)
        good_responses = eliminate_losing_moves(next_state, opponent)
        if not good_responses:
            return candidate_move
    return None
```

Calculates what the board would look like if you play this move

No matter what move you pick, your opponent can prevent a win.

Does your opponent have a good defense? If not, pick this move.

Your opponent will anticipate that you'll try to do this, and also try to block such a play. You can start to see a general strategy forming:

1 See if you can win on the next move. If so, play that move.
2 If not, see if your opponent can win on the next move. If so, block that.
3 If not, see if you can force a win in two moves. If so, play to set that up.
4 If not, see if your opponent could set up a two-move win on their next move.

Notice that all three of your functions have a similar structure. Each function loops over all valid moves and examines the hypothetical board position that you'd get after playing that move. Furthermore, each function builds on the previous function to simulate what your opponent would do in response. If you generalize this concept, you get an algorithm that can always identify the best possible move.

4.3   *Solving tic-tac-toe: a minimax example*

In the previous section, you examined how to anticipate your opponent's play one or two moves ahead. In this section, we show how to generalize that strategy to pick perfect moves in tic-tac-toe. The core idea is exactly the same, but you need the flexibility to look an arbitrary number of moves in the future.

First let's define an enum that represents the three possible outcomes of a game: win, loss, or draw. These possibilities are defined relative to a particular player: a loss for one player is a win for the other.

Listing 4.4   An enum to represent the outcome of a game

```
class GameResult(enum.Enum):
    loss = 1
    draw = 2
    win = 3
```

Imagine you have a function best_result that takes a game state and tells you the best outcome that a player could achieve from that state. If that player could guarantee a win—by any sequence, no matter how complicated—the best_result function

would return GameResult.win. If that player could force a draw, the function would
return GameResult.draw. Otherwise, it would return GameResult.loss. If you assume
that function already exists, it's easy to write a function to pick a move: you loop over
all possible moves, call best_result, and pick the move that leads to the best result
for you. Multiple moves might exist that lead to equal results; you can just pick randomly from them in that case. The following listing shows how to implement this.

Listing 4.5   A game-playing agent that implements minimax search

```
class MinimaxAgent(Agent):
    def select_move(self, game_state):
        winning_moves = []
        draw_moves = []                          Loops over all legal moves
        losing_moves = []                                                       Calculates the
        for possible_move in game_state.legal_moves():     ◄────              game state if you
            next_state = game_state.apply_move(possible_move)  ◄──           select this move
            opponent_best_outcome = best_result(next_state)
            our_best_outcome = reverse_game_result(opponent_best_outcome)
            if our_best_outcome == GameResult.win:
                winning_moves.append(possible_move)
            elif our_best_outcome == GameResult.draw:        Categorizes this move
                draw_moves.append(possible_move)             according to its outcome
            else:
                losing_moves.append(possible_move)
        if winning_moves:
            return random.choice(winning_moves)
        if draw_moves:                                   Picks a move that leads
            return random.choice(draw_moves)             to your best outcome
        return random.choice(losing_moves)
```

Because your opponent plays next, figure out their best possible outcome from there. Your outcome is the opposite of that.

Now the question is how to implement best_result. As in the previous section, you can
start from the end of the game and work backward. The following listing shows the easy
case: if the game is already over, there's only one possible result. You just return it.

Listing 4.6   First step of the minimax search algorithm

```
def best_result(game_state):
    if game_state.is_over():
        if game_state.winner() == game_state.next_player:
            return GameResult.win
        elif game_state.winner() is None:
            return GameResult.draw
        else:
            return GameResult.loss
```

If you're somewhere in the middle of the game, you need to search ahead. By now, the
pattern should be familiar. You start by looping over all possible moves and calculating
the next game state. Then you must assume that your opponent will do their best to
counter your hypothetical move. To do so, you can call best_result from this new
position. That tells you the result that your *opponent* can get from the new position;

you invert it to find out your result. Out of all the moves you consider, you select the one that leads to the best result for you. Listing 4.7 shows how to implement this logic, which makes up the second half of `best_result`. Figure 4.5 illustrates the board positions this function will consider for a particular tic-tac-toe board.

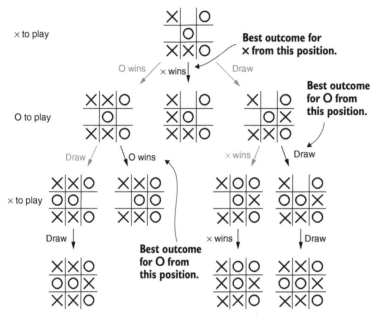

Figure 4.5   A tic-tac-toe game tree. In the top position, it's ×'s turn. If × plays in the top center, O can guarantee a win. If × plays in the left center, × will win. If × plays in the right center, O can force a draw. Therefore, × will choose to play in the left center.

Listing 4.7   Implementing minimax search

Find out your opponent's best move.

See what the board would look like if you play this move.

```
best_result_so_far = GameResult.loss
for candidate_move in game_state.legal_moves():
    next_state = game_state.apply_move(candidate_move)
    opponent_best_result = best_result(next_state)
    our_result = reverse_game_result(opponent_best_result)
    if our_result.value > best_result_so_far.value:
        best_result_so_far = our_result
return best_result_so_far
```

Whatever your opponent wants, you want the opposite.

See if this result is better than the best you've seen so far.

If you apply this algorithm to a simple game such as tic-tac-toe, you get an unbeatable opponent. You can play against it and see for yourself: try the play_ttt.py example on GitHub (http://mng.bz/gYPe). In theory, this algorithm would also work for chess, Go, or any other deterministic, perfect information game. In reality, this algorithm is far too slow for any of those games.

4.4 *Reducing search space with pruning*

In our tic-tac-toe game, you calculated every single possible game in order to find the perfect strategy. There are fewer than 300,000 possible tic-tac-toe games, peanuts for a modern computer. Can you apply the same technique to more interesting games? There are around 500 billion billion (that's a 5 followed by 20 zeros) possible board positions in checkers, for example. It's technically possible to search them all on a cluster of modern computers, but it takes years. In chess and Go, there are more possible board positions than there are atoms in the universe (as their fans are quick to point out). Searching them all is out of the question.

To use tree search to play a sophisticated game, you need a strategy to eliminate parts of the tree. Identifying which parts of the tree you can skip is called *pruning*.

Game trees are two-dimensional: they have width and depth. The *width* is the number of possible moves from a given board position. The *depth* is the number of turns from a board position to a final game state—a possible end of the game. Within a game, both these quantities vary from turn to turn.

You generally estimate the size of the tree by thinking of the typical width and typical depth for a particular game. The number of board positions in a game tree is roughly given by the formula W^d, where W is the average width and d is the average depth. Figures 4.6 and 4.7 illustrate the width and depth of a tic-tac-toe game tree. For example, in chess, a player normally has about 30 options per move, and a game lasts about 80 moves; the size of the tree is something like $30^{80} \approx 10^{118}$ positions. Go typically has 250 legal moves per turn, and a game might last 150 turns. This gives a game tree size of $250^{150} \approx 10^{359}$ positions.

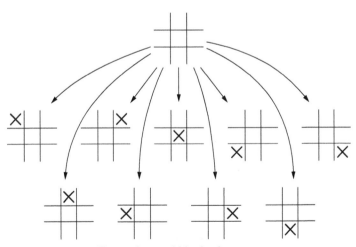

The maximum width of a tic-tac-toe game tree is 9 (on the first move of the game).

Figure 4.6 The width of a tic-tac-toe game tree: the maximum width is 9, because you have 9 possible options on the first move. But the number of legal moves decreases on each turn, so the average width is 4 or 5 moves.

This formula, W^d, is an example of exponential growth: the number of positions to consider grows quickly as you increase the search depth. Imagine a game with an average width and depth of about 10. The full game tree would contain 10^{10}, or 10 billion, board positions to search.

Now suppose you come up with modest pruning schemes. First, you figure out how to quickly eliminate two moves from consideration on each turn, reducing the effective width to 8. Second, you decide you can figure out the game result by looking just 9 moves ahead instead of 10. This leaves you 8^9 positions to search, or around 130 million. Compared to a full search, you've eliminated more than 98% of the computation! The key takeaway is that even slightly reducing the width or depth of your search can massively reduce the time required to select a move. Figure 4.8 illustrates the impact of pruning on a small tree.

In this section, we cover two pruning techniques: *position evaluation functions* for reducing the search depth, and *alpha-beta pruning* for reducing the search width. The two techniques work together to form the backbone of classical board-game AI.

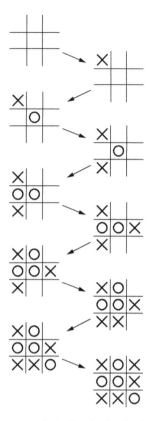

Figure 4.7   The depth of a tic-tac-toe game tree: the maximum depth is 9 moves; after that, the board is full.

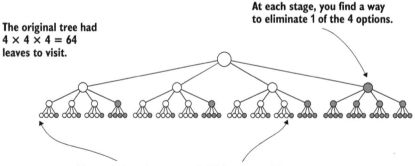

The original tree had $4 \times 4 \times 4 = 64$ leaves to visit.

At each stage, you find a way to eliminate 1 of the 4 options.

After pruning, there are only 27 leaves to visit.

Figure 4.8   Pruning can quickly shrink a game tree. This tree has width 4 and height 3, for a total of 64 leaves to examine. Suppose you find a way to eliminate 1 of the 4 possible options on each turn. Then you end up with just 27 leaves to visit.

4.4.1 Reducing search depth with position evaluation

If you follow a game tree all the way to the end of the game, you can calculate the winner. What about earlier in the game? Human players normally have a sense of who is leading throughout the midgame. Even beginner Go players have an instinctive feel for whether they're bossing their opponent around or scrambling to survive. If you can capture that sense in a computer program, you can reduce the depth that you need to search. A function that mimics this sense of who is leading, and by how much, is a *position evaluation function.*

For many games, the position evaluation function can be handcrafted by using knowledge of the game. For example:

- *Checkers*—Count one point for each regular piece on the board, plus two points for each king. Take the value of your pieces and subtract the value of your opponent's.
- *Chess*—Count one point for each pawn, three points for each bishop or knight, five points for each rook, and nine points for the queen. Take the value of your pieces and subtract the value of your opponent's.

These evaluation functions are highly simplified; top checkers and chess engines use much more sophisticated heuristics. But in both cases, the AI will have the incentive to capture the opponent's pieces and preserve its own. Furthermore, it'll be willing to trade its weaker pieces to capture a stronger one.

In Go, the equivalent heuristic is to add up the stones you've captured and then subtract the number of stones your opponent has captured. (Equivalently, you can count the difference in the number of stones on the board.) Listing 4.8 calculates that heuristic. It turns out this isn't an effective evaluation function. In Go, the *threat* of capturing stones is much more important than *actually* capturing them. It's quite common for a game to last for 100 turns or more before any stones are captured. Crafting a board evaluation function that accurately captures the nuances of the game state turns out to be extremely difficult.

That said, you can use this too-simple heuristic for the purpose of illustrating pruning techniques. This isn't going to produce a strong bot, but it's better than picking moves completely at random. In chapters 11 and 12, we cover how to use deep learning to generate a better evaluation function.

After you've chosen an evaluation function, you can implement *depth pruning.* Instead of searching all the way to the end of the game and seeing who won, you search a fixed number of moves ahead and use the evaluation function to estimate who is more likely to win.

Listing 4.8 A highly simplified board evaluation heuristic for Go

```
def capture_diff(game_state):
    black_stones = 0
    white_stones = 0
    for r in range(1, game_state.board.num_rows + 1):
```

```
            for c in range(1, game_state.board.num_cols + 1):
                p = gotypes.Point(r, c)
                color = game_state.board.get(p)
                if color == gotypes.Player.black:
                    black_stones += 1
                elif color == gotypes.Player.white:
                    white_stones += 1
        diff = black_stones - white_stones
        if game_state.next_player == gotypes.Player.black:
            return diff
        return -1 * diff
```

If it's black's move, return (black stones) – (white stones).

Calculate the difference between the number of black stones and white stones on the board. This will be the same as the difference in the number of captures, unless one player passes early. → `diff = black_stones - white_stones`

If it's white's move, return (white stones) – (black stones). ←── `return -1 * diff`

Figure 4.9 shows a partial game tree with depth pruning. (We've left most of the branches out of the diagram to save space, but the algorithm would examine those too.)

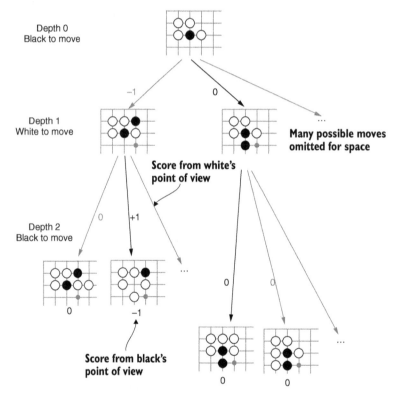

Figure 4.9 A partial Go game tree. Here you search the tree to a depth of 2 moves ahead. At that point, you look at the number of captures to evaluate the board. If black chooses the rightmost branch, white can capture a stone, yielding an evaluation of –1 for black. If black chooses the center branch, the black stone is safe (for now). That branch is evaluated at a score of 0. Therefore, black chooses the center branch.

In this tree, you're searching to a depth of 2 moves ahead, and using the number of captured stones as the board evaluation function. The original position shows black to play; black has a stone with just one liberty. What should black do? If black extends straight down, as shown in the middle branch, the stone is safe (for now). If black plays anywhere else, white can capture the stone—the left branch shows one of the many ways that can happen.

After looking two moves ahead, you apply your evaluation function to the position. In this case, any branch where white captures the stone is scored at +1 for white and –1 for black. All other branches have a score of 0 (there's no other way to capture a stone in two turns). In this case, black picks the only move that protects the stone.

Listing 4.9 shows how to implement depth pruning. The code looks similar to the full minimax code in listing 4.7: it may be helpful to compare them side by side. Note the differences:

- Instead of returning a win/lose/draw enum, you return a number indicating the value of your board evaluation function. Our convention is that the score is from the perspective of the player who has the next turn: a large score means the player who has the next move expects to win. When you evaluate the board from your opponent's perspective, you multiply the score by –1 to flip back to your perspective.
- The max_depth parameter controls the number of moves you want to search ahead. At each turn, you subtract 1 from this value.
- When max_depth hits 0, you stop searching and call your board evaluation function.

Listing 4.9 Depth-pruned minimax search

```
def best_result(game_state, max_depth, eval_fn):
    if game_state.is_over():
        if game_state.winner() == game_state.next_player:
            return MAX_SCORE
        else:
            return MIN_SCORE

    if max_depth == 0:
        return eval_fn(game_state)

    best_so_far = MIN_SCORE
    for candidate_move in game_state.legal_moves():
        next_state = game_state.apply_move(candidate_move)
        opponent_best_result = best_result(
            next_state, max_depth - 1, eval_fn)
        our_result = -1 * opponent_best_result
        if our_result > best_so_far:
            best_so_far = our_result

    return best_so_far
```

If the game is already over, you know who the winner is.

Loop over all possible moves.

You've reached your maximum search depth. Use your heuristic to decide how good this sequence is.

See what the board would look like if you play this move.

Find the opponent's best result from this position.

Whatever your opponent wants, you want the opposite.

See if this is better than the best result you've seen so far.

Feel free to experiment with your own evaluation functions. It can be fun to see how they affect your bot's personality, and you can certainly do a bit better than our simple example.

4.4.2   Reducing search width with alpha-beta pruning

Look at the diagram in figure 4.10. It's black's move, and you're considering playing on the point marked with a square. If you do so, white can then play at A to capture four stones. Clearly that's a disaster for black! What if white replies at B instead? Well, who cares? White's response at A is bad enough already. From black's perspective, you don't really care if A is the absolute best move white can pick. As soon as you find one powerful response, you can reject playing at the point marked with a square and move on to the next option. That's the idea behind *alpha-beta pruning*.

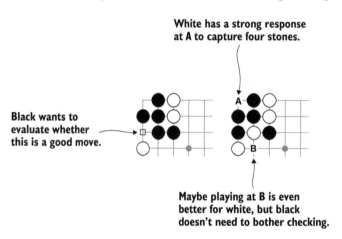

Figure 4.10   The black player is considering playing at the point marked with a square. If black plays there, white can respond at A to capture four stones. That result is so bad for black that you can immediately reject playing on the square; there's no need for black to consider other white responses, such as B.

Let's walk through how the alpha-beta algorithm would apply to that position. Alpha-beta pruning starts like a regular depth-pruned tree search. Figure 4.11 shows the first step. You pick the first move to evaluate for black; that move is marked with an A in the diagram. Then you fully evaluate that move up to a depth of 3. You can see that no matter how white responds, black can capture at least two stones. So you evaluate this branch to a score of +2 for black.

Now you consider the next candidate move for black, marked as B in figure 4.12. Just as in the depth-pruned search, you look at all possible white responses and evaluate them one by one. White can play in the upper-left corner to capture four stones; that branch gets evaluated at a score of –4 for black. Now, you already know that if black plays at A, black is guaranteed a score of at least +2. If black plays at B, you just

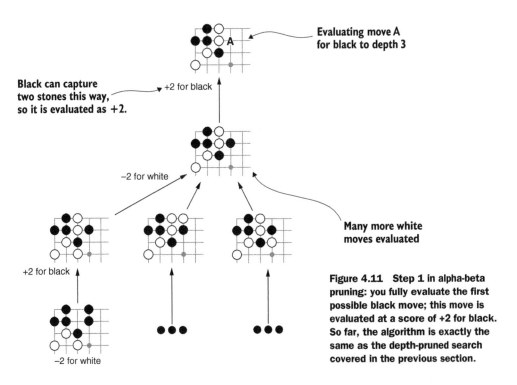

Evaluating move A for black to depth 3

Black can capture two stones this way, so it is evaluated as +2.

+2 for black

−2 for white

+2 for black

−2 for white

Many more white moves evaluated

Figure 4.11 Step 1 in alpha-beta pruning: you fully evaluate the first possible black move; this move is evaluated at a score of +2 for black. So far, the algorithm is exactly the same as the depth-pruned search covered in the previous section.

saw how white can hold black to a score of –4; possibly white can do even better. But because –4 is already worse than +2, there's no need to search further. You can skip evaluating a dozen other white responses—and each of those positions has many more combinations after it. The amount of computation you save adds up quickly, and you still select the exact same move that you would've chosen with a full search to depth 3.

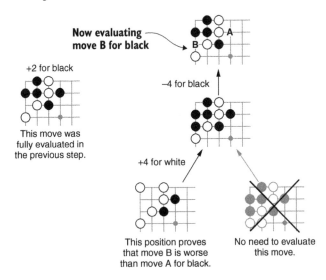

Now evaluating move B for black

+2 for black

This move was fully evaluated in the previous step.

−4 for black

+4 for white

This position proves that move B is worse than move A for black.

No need to evaluate this move.

Figure 4.12 Step 2 in alpha-beta pruning: you now evaluate the second possible black move. Here, white has a response that captures four stones. That branch evaluates to –4 for black. As soon as you evaluate that white response, you can discard this move for black entirely and skip the other possible white responses. It's possible that white has an even better reply that you didn't evaluate, but all you need to know is that playing at B is worse for black than playing at A.

For the purposes of this example, we chose a specific order to evaluate the moves to illustrate how the pruning works. Our actual implementation evaluates moves in the order of their board coordinates. The time savings you get by alpha-beta pruning depends on how quickly you find good branches. If you happen to evaluate the best branches early on, you can eliminate the other branches quickly. In the worst case, you evaluate the best branch last, and then alpha-beta pruning is no faster than a full depth-pruned search.

To implement the algorithm, you must track the best result so far for each player throughout your search. These values are traditionally called *alpha* and *beta*, which is where the name of the algorithm comes from. In our implementation, we call those values best_black and best_white.

Listing 4.10 Checking whether you can stop evaluating a branch

```
def alpha_beta_result(game_state, max_depth,
                      best_black, best_white, eval_fn):
    ...
        if game_state.next_player == Player.white:
            # Update our benchmark for white.
            if best_so_far > best_white:
                best_white = best_so_far
            outcome_for_black = -1 * best_so_far
            if outcome_for_black < best_black:
                return best_so_far
```

Updates your benchmark for white

You're picking a move for white; it needs to be only strong enough to eliminate black's previous move. As soon as you find something that defeats black's best option, you can stop searching.

You can extend your implementation of depth pruning to include alpha-beta pruning as well. Listing 4.10 shows the key new addition. This block is implemented from white's perspective; you need a similar block for black.

First, you check whether you need to update the best_white score. Next, you check whether you can stop evaluating moves for white. You do this by comparing the current score to the best score you've found for black in *any* branch. If white can hold black to a lower score, black won't choose this branch; you don't need to find the absolute best score.

The full implementation of alpha-beta pruning is shown in the following listing.

Listing 4.11 Full implementation of alpha-beta pruning

```
def alpha_beta_result(game_state, max_depth,
                      best_black, best_white, eval_fn):
    if game_state.is_over():
        if game_state.winner() == game_state.next_player:
            return MAX_SCORE
        else:
            return MIN_SCORE

    if max_depth == 0:
        return eval_fn(game_state)

    best_so_far = MIN_SCORE
    for candidate_move in game_state.legal_moves():
        next_state = game_state.apply_move(candidate_move)
```

Check if the game is already over.

You've reached your maximum search depth. Use your heuristic to decide how good this sequence is.

Loop over all valid moves.

See what the board would look like if you play this move.

Whatever your opponent wants, you want the opposite.

Find out your opponent's best result from that position.

```
opponent_best_result = alpha_beta_result(
    next_state, max_depth - 1,
    best_black, best_white,
    eval_fn)
our_result = -1 * opponent_best_result
```

Update your benchmark for white.

See whether this result is better than the best you've seen so far.

```
if our_result > best_so_far:
    best_so_far = our_result
if game_state.next_player == Player.white:
    if best_so_far > best_white:
        best_white = best_so_far
    outcome_for_black = -1 * best_so_far
    if outcome_for_black < best_black:
        return best_so_far
elif game_state.next_player == Player.black:
    if best_so_far > best_black:
        best_black = best_so_far
    outcome_for_white = -1 * best_so_far
    if outcome_for_white < best_white:
        return best_so_far
```

Update your benchmark for black.

You're picking a move for white; it needs to be only strong enough to eliminate black's previous move.

You're picking a move for black; it needs to be only strong enough to eliminate white's previous move.

```
return best_so_far
```

4.5 Evaluating game states with Monte Carlo tree search

For alpha-beta pruning, you used a position evaluation function to help reduce the number of positions you had to consider. But position evaluation in Go is very, very hard: your simple heuristic based on captures won't fool many Go players. *Monte Carlo tree search* (MCTS) provides a way to evaluate a game state without *any* strategic knowledge about the game. Instead of a game-specific heuristic, the MCTS algorithm simulates random games to estimate how good a position is. One of these random games is called a *rollout* or a *playout*. In this book, we use the term *rollout*.

Monte Carlo tree search is part of the larger family of *Monte Carlo algorithms*, which use randomness to analyze extremely complex situations. The element of randomness inspired the name, an allusion to the famous casino district in Monaco.

It may seem impossible that you can build a good strategy out of picking random moves. A game AI that chooses completely random moves is going to be extremely weak, of course. But when you pit two random AIs against each other, the opponent is equally clueless. If black consistently wins more often than white, it must be because black started with an advantage. Therefore, you can figure out whether a position gives one player an advantage by starting random games from there. And you don't need any understanding of *why* the position is good to do this.

It's possible to get unbalanced results by chance. If you simulate 10 random games and white wins seven, how confident can you be that white had an advantage? Not very: white has won only two more games than you'd expect by chance. If black and white were perfectly balanced, there's about a 30% chance of seeing a 7-out-of-10 result. On the other hand, if white wins 70 out of 100 random games, you can be virtually certain that the starting position did favor white. The key idea is that your estimate gets more accurate as you do more rollouts.

Each round of the MCTS algorithm takes three steps:

1 Add a new board position to the MCTS tree.
2 Simulate a random game from that position.
3 Update the tree statistics with the results of that random game.

You repeat this process as many times as you can in the time available. Then the statistics at the top of the tree tell you which move to pick.

Let's step through a single round of the MCTS algorithm. Figure 4.13 shows an MCTS tree. At this point in the algorithm, you've already completed a number of rollouts and built up a partial tree. Each node tracks the counts of who won the rollouts that started from any board position after that node. Every node's count includes the sum of all its children. (Normally, the tree would have many more nodes at this point; in the diagram, we've omitted many of the nodes to save space.)

At each round, you add a new game position to the tree. First, you pick a node at the bottom of the tree (a *leaf*) where you want to add a new child. This tree has five leaves.

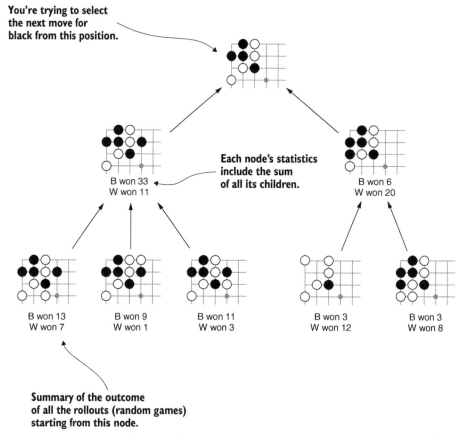

You're trying to select the next move for black from this position.

Each node's statistics include the sum of all its children.

B won 33
W won 11

B won 6
W won 20

B won 13
W won 7

B won 9
W won 1

B won 11
W won 3

B won 3
W won 12

B won 3
W won 8

Summary of the outcome of all the rollouts (random games) starting from this node.

Figure 4.13 An MCTS game tree. The top of the tree represents the current board position; you're trying to find the next move for black. At this point in the algorithm, you've performed 70 random rollouts from various possible positions. Each node tracks statistics on all the rollouts that started from any of its children.

To get the best results, you need to be a little careful about the way you pick a leaf; section 4.5.2 covers a good strategy for doing so. For now, just suppose that you walked all the way down the leftmost branches. From that point, you randomly pick the next move, calculate the new board position, and add that node to the tree. Figure 4.14 shows what the tree looks like after that process.

The new node in the tree is the starting point for a random game. You simulate the rest of the game, literally just selecting any legal play at each turn, until the game is over. Then you count up the score and find the winner. In this case, let's suppose the winner is white. You record the result of this rollout in the new node. In addition, you walk up to all the node's ancestors and add the new rollout to their counts as well. Figure 4.15 shows what the tree looks like after this step is complete.

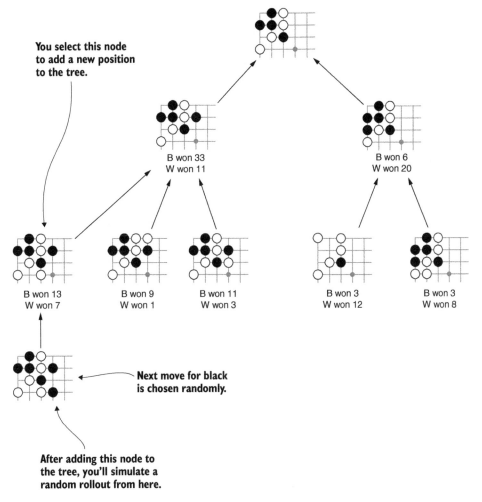

Figure 4.14 Adding a new node to an MCTS tree. Here you select the leftmost branch as the place to insert a new node. You then randomly select the next move from the position to create the new node in the tree.

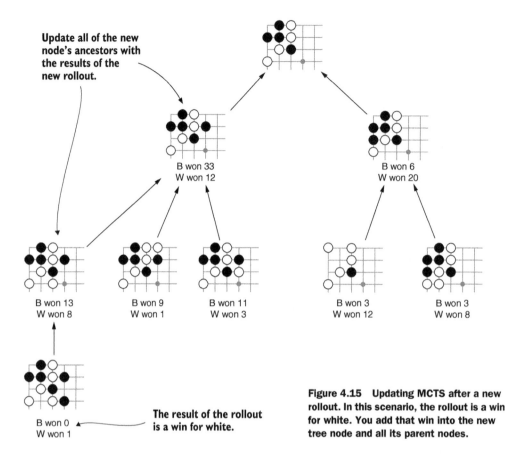

Update all of the new node's ancestors with the results of the new rollout.

B won 33
W won 12

B won 6
W won 20

B won 13
W won 8

B won 9
W won 1

B won 11
W won 3

B won 3
W won 12

B won 3
W won 8

B won 0
W won 1

The result of the rollout is a win for white.

Figure 4.15   Updating MCTS after a new rollout. In this scenario, the rollout is a win for white. You add that win into the new tree node and all its parent nodes.

That whole process is a single round of MCTS. Every time you repeat it, the tree gets bigger, and the estimates at the top get more accurate. Normally, you stop after a fixed number of rounds or after a fixed amount of time passes. At that point, you select the move that has the highest winning percentage.

4.5.1   Implementing Monte Carlo tree search in Python

Now that you've walked through the MCTS algorithm, let's look at the implementation details. First, you'll design a data structure to represent the MCTS tree. Next, you'll write a function to execute the MCTS rollouts.

As shown in listing 4.12, you start by defining a new class, MCTSNode, to represent any node in your tree. Each MCTSNode will track the following properties:

- game_state—The current state of the game (board position and next player) at this node in the tree.
- parent—The parent MCTSNode that led to this one. You can set parent to None to indicate the root of the tree.
- move—The last move that directly led to this node.

- children—A list of all child nodes in the tree.
- win_counts and num_rollouts—Statistics about the rollouts that started from this node.
- unvisited_moves—A list of all legal moves from this position that aren't yet part of the tree. Whenever you add a new node to the tree, you pull one move out of unvisited_moves, generate a new MCTSNode for it, and add it to the children list.

Listing 4.12 A data structure to represent an MCTS tree

```
class MCTSNode(object):
    def __init__(self, game_state, parent=None, move=None):
        self.game_state = game_state
        self.parent = parent
        self.move = move
        self.win_counts = {
            Player.black: 0,
            Player.white: 0,
        }
        self.num_rollouts = 0
        self.children = []
        self.unvisited_moves = game_state.legal_moves()
```

An MCTSNode can be modified in two ways. You can add a new child to the tree, and you can update its rollout stats. The following listing shows both functions.

Listing 4.13 Methods to update a node in an MCTS tree

```
    def add_random_child(self):
        index = random.randint(0, len(self.unvisited_moves) - 1)
        new_move = self.unvisited_moves.pop(index)
        new_game_state = self.game_state.apply_move(new_move)
        new_node = MCTSNode(new_game_state, self, new_move)
        self.children.append(new_node)
        return new_node

    def record_win(self, winner):
        self.win_counts[winner] += 1
        self.num_rollouts += 1
```

Finally, you add three convenience methods to access useful properties of your tree node:

- can_add_child reports whether this position has any legal moves that haven't yet been added to the tree.
- is_terminal reports whether the game is over at this node; if so, you can't search any further from here.
- winning_frac returns the fraction of rollouts that were won by a given player.

These functions are implemented in the following listing.

Listing 4.14   Helper methods to access useful MCTS tree properties

```
def can_add_child(self):
    return len(self.unvisited_moves) > 0

def is_terminal(self):
    return self.game_state.is_over()

def winning_frac(self, player):
    return float(self.win_counts[player]) / float(self.num_rollouts)
```

Having defined the data structure for the tree, you can now implement the MCTS algorithm. You start by creating a new tree. The root node is the current game state. Then you repeatedly generate rollouts. In this implementation, you loop for a fixed number of rounds for each turn; other implementations may run for a specific length of time instead.

Each round begins by walking down the tree until you find a node where you can add a child (any board position that has a legal move that isn't yet in the tree). The select_move function hides the work of choosing the best branch to explore; we cover the details in the next section.

After you find a suitable node, you call add_random_child to pick any follow-up move and bring it into the tree. At this point, node is a newly created MCTSNode that has zero rollouts.

You now start a rollout from this node by calling simulate_random_game. The implementation of simulate_random_game is identical to the bot_v_bot example covered in chapter 3.

Finally, you update the win counts of the newly created node and all its ancestors. This whole process is implemented in the following listing.

Listing 4.15   The MCTS algorithm

```
class MCTSAgent(agent.Agent):
    def select_move(self, game_state):
        root = MCTSNode(game_state)

        for i in range(self.num_rounds):
            node = root
            while (not node.can_add_child()) and (not node.is_terminal()):
                node = self.select_child(node)

            if node.can_add_child():                          # Adds a new child
                node = node.add_random_child()                # node into the tree

            winner = self.simulate_random_game(node.game_state)  # Simulates a random game from this node

            while node is not None:
                node.record_win(winner)                       # Propagates the score
                node = node.parent                            # back up the tree
```

After you've completed the allotted rounds, you need to pick a move. To do so, you just loop over all the top-level branches and pick the one with the best winning percentage. The following listing shows how to implement this.

Listing 4.16 Selecting a move after completing your MCTS rollouts

```
class MCTSAgent:
...
    def select_move(self, game_state):
...
        best_move = None
        best_pct = -1.0
        for child in root.children:
            child_pct = child.winning_pct(game_state.next_player)
            if child_pct > best_pct:
                best_pct = child_pct
                best_move = child.move
        return best_move
```

4.5.2 *How to select which branch to explore*

Your game AI has a limited amount of time to spend on each turn, which means you can perform only a fixed number of rollouts. Each rollout improves your evaluation of a single possible move. Think of your rollouts as a limited resource: if you spend an extra rollout on move A, you have to spend one fewer rollout on move B. You need a strategy to decide how to allocate your limited budget. The standard strategy is called the *upper confidence bound for trees*, or *UCT* formula. The UCT formula strikes a balance between two conflicting goals.

The first goal is to spend your time looking at the best moves. This goal is called *exploitation* (you want to exploit any advantage that you've discovered so far). You'd spend more rollouts on the moves with the highest estimated winning percentage. Now, some of those moves have a high winning percentage just by chance. But as you complete more rollouts through those branches, your estimates get more accurate. The false positives will drop lower down the list.

On the other hand, if you've visited a node only a few times, your estimate may be way off. By sheer chance, you may have a low estimate for a move that's really good. Spending a few more rollouts there may reveal its true quality. So your second goal is to get more accurate evaluations for the branches you've visited the least. This goal is called *exploration*.

Figure 4.16 compares a search tree biased toward exploitation against a tree biased toward exploration. The exploitation-exploration trade-off is a common feature of trial-and-error algorithms. It'll come up again when we look at reinforcement learning later in the book.

For each node you're considering, you compute the winning percentage w to represent the exploitation goal. To represent exploration, you compute, where N is the total number of rollouts, and n is the number of rollouts that started with the node under consideration. This specific formula has a theoretical basis; for our purposes, just note that its value will be largest for the nodes you've visited the least.

$$\sqrt{\frac{\log N}{n}}$$

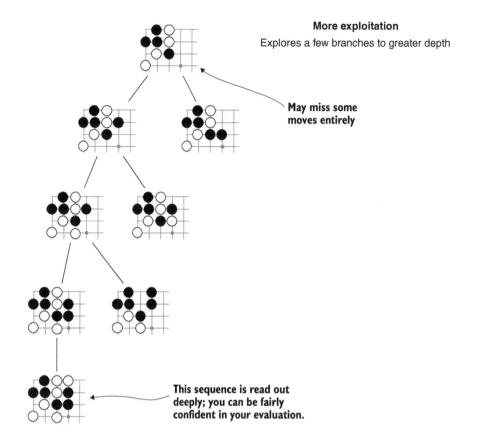

More exploitation

Explores a few branches to greater depth

May miss some moves entirely

This sequence is read out deeply; you can be fairly confident in your evaluation.

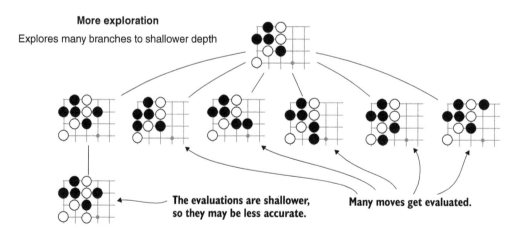

More exploration

Explores many branches to shallower depth

The evaluations are shallower, so they may be less accurate.

Many moves get evaluated.

Figure 4.16   The exploitation-exploration trade-off. In both game trees, you've visited seven board positions. On top, the search is biased more toward exploitation: the tree is deeper for the most promising moves. On the bottom, the search is biased more toward exploration: it tries more moves, but to less depth.

You combine these two components to get the UCT formula:

$$w + c\sqrt{\frac{\log N}{n}}$$

Here, c is a parameter that represents your preferred balance between exploitation and exploration. The UCT formula gives you a score for each node, and the node with the highest UCT score is the start of the next rollout.

With a larger value of c, you'll spend more time visiting the least-explored nodes. With a smaller value of c, you'll spend more time gathering a better evaluation of the most promising node. The choice of c that makes the most effective game player is usually found via trial and error. We suggest starting somewhere around 1.5 and experimenting from there. The parameter c is sometimes called the *temperature*. When the temperature is "hotter," your search will be more volatile, and when the temperature is "cooler," your search will be more focused.

Listing 4.17 shows how to implement this policy. After you've identified the metric you want to use, selecting a child is a simple matter of calculating the formula for each node and choosing the node with the largest value. Just as in minimax search, you need to switch your perspective on each turn. You calculate the winning percentage from the point of view of the player who'd pick the next move, so that perspective alternates between black and white as you walk down the tree.

Listing 4.17 Selecting a branch to explore with the UCT formula

```python
def uct_score(parent_rollouts, child_rollouts, win_pct, temperature):
    exploration = math.sqrt(math.log(parent_rollouts) / child_rollouts)
    return win_pct + temperature * exploration

class MCTSAgent:
...
    def select_child(self, node):
        total_rollouts = sum(child.num_rollouts for child in node.children)

        best_score = -1
        best_child = None
        for child in node.children:
            score = uct_score(
                total_rollouts,
                child.num_rollouts,
                child.winning_pct(node.game_state.next_player),
                self.temperature)
            if score > best_score:
                best_score = uct_score
                best_child = child
        return best_child
```

4.5.3   *Applying Monte Carlo tree search to Go*

In the previous section, you implemented the general form of the MCTS algorithm. Straightforward MCTS implementations can reach the amateur 1 dan level for Go, the level of a strong amateur player. Combining MCTS with other techniques can produce a bot that's a fair bit stronger than that; many of the top Go AIs today use both MCTS and deep learning. If you're interested in reaching a competitive level with your MCTS bot, this section covers some of the practical details to consider.

FAST CODE MAKES A STRONG BOT

MCTS starts to be a viable strategy for full-size (19 × 19) Go at around 10,000 rollouts per turn. The implementation in this chapter isn't fast enough to do that: you'll be waiting several minutes for it to choose each move. You'll need to optimize your implementation a bit in order to complete that many rollouts in a reasonable amount of time. On small boards, on the other hand, even your reference implementation makes a fun opponent.

All else being equal, more rollouts means a better decision. You can always make your bot stronger just by speeding up the code so as to squeeze more rollouts in the same amount of time. It's not just the MCTS-specific code that's relevant. The code that calculates captures, for example, is called hundreds of times per rollout. All the basic game logic is fair game for optimization.

BETTER ROLLOUT POLICIES MAKE BETTER EVALUATIONS

The algorithm for selecting moves during random rollouts is called the *rollout policy*. The more realistic your rollout policy is, the more accurate your evaluations will be. In chapter 3, you implemented a RandomAgent that plays Go; in this chapter, you used your RandomAgent as your rollout policy. But it's not quite true that the RandomAgent chooses moves *completely* randomly with no Go knowledge at all. First, you programmed it not to pass or resign before the board is full. Second, you programmed it not to fill its own eyes, so it wouldn't kill its own stones at the end of the game. Without this logic, the rollouts would be less accurate.

Some MCTS implementations go further and implement more Go-specific logic in their rollout policy. Rollouts with game-specific logic are sometimes called *heavy* rollouts; by contrast, rollouts that are close to purely random are sometimes called *light* rollouts.

One way to implement heavy rollouts is to build up a list of basic tactical shapes that are common in Go, along with a known response. Anywhere you find a known shape on the board, you look up the known response and boost its probability of being selected. You don't want to *always* pick the known response as a hard-and-fast rule; that will remove the vital element of randomness from the algorithm.

One example is in figure 4.17. This is a 3 × 3 local pattern in which a black stone is in danger of getting captured on white's next turn. Black can save it, at least temporarily, by extending. This isn't always the best move; it's not even always a good move. But it's more likely to be a good move than any random point on the board.

Pattern	Action
Look for this shape anywhere on the board.	**Anywhere you see that shape, consider this move.**

Figure 4.17 An example of a local tactical pattern. When you see the shape on the left, you should consider the response on the right. A policy that follows tactical patterns like this one won't be especially strong, but will be much stronger than choosing moves completely at random.

Building up a good list of these patterns requires some knowledge of Go tactics. If you're curious about other tactical patterns that you can use in heavy rollouts, we suggest looking at the source code for Fuego (http://fuego.sourceforge.net/) or Pachi (https://github.com/pasky/pachi), two open source MCTS Go engines.

Be careful when implementing heavy rollouts. If the logic in your rollout policy is slow to compute, you can't execute as many rollouts. You may end up wiping out the gains of the more sophisticated policy.

A POLITE BOT KNOWS WHEN TO RESIGN

Making a game AI isn't just an exercise in developing the best algorithm. It's also about creating a fun experience for the human opponent. Part of that fun comes from giving the human player the satisfaction of winning. The first Go bot you implemented in this book, the RandomAgent, is maddening to play against. After the human player inevitably pulls ahead, the random bot insists on continuing until the entire board is full. Nothing is stopping the human player from walking away and mentally scoring the game as a win. But this somehow feels unsporting. It's a much better experience if your bot can gracefully resign instead.

You can easily add human-friendly resignation logic on top of a basic MCTS implementation. The MCTS algorithm computes an estimated winning percentage in the process of selecting a move. Within a single turn, you compare these numbers to decide what move to pick. But you can also compare the estimated winning percentage at different points in the same game. If these numbers are dropping, the game is tilting in the human's favor. When the best option carries a sufficiently low winning percentage, say 10%, you can make your bot resign.

4.6 Summary

- Tree-search algorithms evaluate many possible sequences of decisions to find the best one. Tree search comes up in games as well as general optimization problems.
- The variant of tree search that applies to games is *minimax* tree search. In minimax search, you alternate between two players with opposing goals.
- Full minimax tree search is practical in only extremely simple games (for example, tic-tac-toe). To apply it to sophisticated games (such as chess or Go), you need to reduce the size of the tree that you search.

- A *position evaluation function* estimates which player is more likely to win from a given board position. With a good position evaluation function, you don't have to search all the way to the end of a game in order to make a decision. This strategy is called *depth pruning*.

- Alpha-beta pruning reduces the number of moves you need to consider at each turn, making it practical for games as complex as chess. The idea of alpha-beta pruning is intuitive: when evaluating a possible move, if you find a single strong countermove from your opponent, you can immediately drop that move from consideration entirely.

- When you don't have a good position evaluation heuristic, you can sometimes use *Monte Carlo tree search*. This algorithm simulates random games from a particular position and tracks which player wins more often.

Getting started with neural networks

5

This chapter covers

- Introducing the fundamentals of artificial neural networks
- Teaching a network to recognize handwritten digits
- Creating neural networks by composing layers
- Understanding how neural networks learn from data
- Implementing a simple neural network from scratch

This chapter introduces the core notions of artificial neural networks (ANNs), a class of algorithms central to modern-day *deep learning*. The history of artificial neural networks is a surprisingly old one, dating back to the early 1940s. It took many decades for its applications to become vast successes in many areas, but the basic ideas remain in effect.

At the core of ANNs is the idea to take inspiration from neuroscience and model a class of algorithms that works similarly to the way we hypothesize part of the brain functions. In particular, we use the notion of *neurons* as atomic blocks for our artificial networks. Neurons form groups called *layers*, and these layers are *connected* to each other in specific ways to span a *network*. Given input data, neurons can transfer information layer by layer via connections, and we say that they *activate* if the signal is strong enough. In this way, data is propagated through the network until we arrive at the last step, the output layer, from which we get our *predictions*. These predictions can then be compared to the *expected output* to compute the *error* of the prediction, which the network uses to learn and improve future predictions.

Although the brain-inspired architecture analogy is useful at times, we don't want to overstress it here. We do know a lot about the visual cortex of our brain in particular, but the analogy can sometimes be misleading or even harmful. We think it's better to think of ANNs as trying to uncover *the guiding principles of learning in organisms*, just as an airplane makes use of aerodynamics, but doesn't try to copy a bird.

To make things more concrete in this chapter, we provide a basic implementation of a neural network to follow—starting from scratch. You'll apply this network to tackle a problem from *optical character recognition* (OCR); namely, how to let a computer predict which digit is displayed on an image of a handwritten digit.

Each image in our OCR data set is made up of pixels laid out on a grid, and you must analyze spatial relationships between the pixels to figure out what digit it represents. Go, like many other board games, is played on a grid, and you must consider the spatial relationships on the board in order to pick a good move. You might hope that machine-learning techniques for OCR could also apply to games like Go. As it turns out, they do. Chapters 6 through 8 show how to apply these methods to games.

We keep the mathematics relatively low in this chapter. If you aren't familiar with the basics of linear algebra, calculus, and probability theory or need a brief and practical reminder, we suggest that you read appendix A first. Also, the more difficult parts of the learning procedure of a neural network can be found in appendix B. If you know neural networks, but have never implemented one, we suggest that you skip to section 5.5 right away. If you're familiar with implementing networks as well, jump right into chapter 6, in which you'll apply neural networks to predict moves from games generated in chapter 4.

5.1 A simple use case: classifying handwritten digits

Before introducing neural networks in detail, let's start with a concrete use case. Throughout this chapter, you'll build an application that can predict digits from handwritten image data reasonably well, with about 95% accuracy. Notably, you'll do all this by exposing only the pixel values of the images to a neural network; the algorithm will learn to extract relevant information about the structure of digits on its own.

You'll use the Modified National Institute of Standards and Technology (MNIST) data set of handwritten digits to do so, a well-studied data set among machine-learning practitioners and the fruit fly of deep learning.

In this chapter, you'll use the NumPy library for handling low-level mathematical operations. NumPy is the industry standard for machine learning and mathematical computing in Python, and you'll use it throughout the rest of the book. Before trying out any of the code samples in this chapter, you should install NumPy with your preferred package manager. If you use pip, run `pip install numpy` from a shell to install it. If you use Conda, run `conda install numpy`.

5.1.1 The MNIST data set of handwritten digits

The MNIST data set consists of 60,000 images, 28×28 pixels each. A few examples of this data are shown in figure 5.1. To humans, recognizing most of these examples is a trivial task, and you can easily read the examples in the first row as 7, 5, 3, 9, 3, 0, and so on. But in some cases, it's difficult even for humans to understand what the picture represents. For instance, the fourth picture in row five in figure 5.1 could easily be a 4 or a 9.

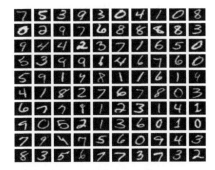

Figure 5.1 A few samples from the MNIST data set for handwritten digits, a well-studied entity in the field of optical character recognition

Each image in MNIST is annotated with a *label*, a digit from 0 to 9 representing the true value depicted on the image.

Before you can look at the data, you need to load it first. In our GitHub repository for this book, you'll find a file called mnist.pkl.gz located in the folder http://mng.bz/P8mn.

In this folder, you'll also find all the code you'll write in this chapter. As before, we suggest that you follow the flow of this chapter and build the code base as you go, but you can also run the code as found in the GitHub repository.

5.1.2 MNIST data preprocessing

Because the labels in this data set are integers from 0 to 9, you'll use a technique called *one-hot encoding* to transform the digit 1 to a vector of length 10 with all 0s, except you place a 1 at position 1. This representation is useful and widely used in the context of machine learning. Reserving the first slot in a vector for label 1 allows algorithms such as neural networks to distinguish more easily between labels. Using one-hot encoding, the digit 2, for instance, has the following representation: [0, 0, 1, 0, 0, 0, 0, 0, 0, 0].

Listing 5.1 One-hot encoding of MNIST labels

```
import six.moves.cPickle as pickle
import gzip
```

```
import numpy as np

def encode_label(j):
    e = np.zeros((10, 1))
    e[j] = 1.0
    return e
```

You one-hot encode indices to vectors of length 10.

The benefit of one-hot encoding is that each digit has its own "slot," and you can use neural networks to output *probabilities* for an input image, which will become useful later.

Examining the contents of the file mnist.pkl.gz, you have access to three pools of data: training, validation, and test data. Recall from chapter 1 that you use training data to train, or fit, a machine-learning algorithm, and use test data to evaluate how well the algorithm learned. Validation data can be used to tweak and validate the configuration of the algorithm, but can be safely ignored in this chapter.

Images in the MNIST data set are quadratic and have both height and width of 28 pixels. You load image data into *feature vectors* of size $784 = 28 \times 28$; you discard the image structure altogether and look only at pixels represented as a vector. Each value of this vector represents a grayscale value between 0 and 1, with 0 being white and 1 black.

Listing 5.2   Reshaping MNIST data and loading training and test data

```
def shape_data(data):
    features = [np.reshape(x, (784, 1)) for x in data[0]]

    labels = [encode_label(y) for y in data[1]]

    return zip(features, labels)


def load_data():
    with gzip.open('mnist.pkl.gz', 'rb') as f:
        train_data, validation_data, test_data = pickle.load(f)

    return shape_data(train_data), shape_data(test_data)
```

Flatten the input images to feature vectors of length 784.

All labels are one-hot encoded.

Create pairs of features and labels.

Unzipping and loading the MNIST data yields three data sets.

Discard validation data here and reshape the other two data sets.

You now have a simple representation of the MNIST data set; both features and labels are encoded as vectors. Your task is to devise a mechanism that learns to map features to labels accurately. Specifically, you want to design an algorithm that takes in training features and labels to learn, such that it can predict labels for test features.

Neural networks can do this job well, as you'll see in the next section, but let's first discuss a naive approach that will show you the general problems you have to tackle for this application. Recognizing digits is a relatively simple task for humans, but it's difficult to explain exactly how to do it and how we know what we know. This phenomenon of knowing more than you can explain is called *Polanyi's paradox*. This makes it particularly hard to describe to a machine *explicitly* how to solve this problem.

One aspect that plays a crucial role is *pattern recognition*—each handwritten digit has certain traits that derive from its prototypical, digital version. For instance, a 0 is roughly an oval, and in many countries a 1 is simply a vertical line. Given this heuristic, you can naively approach classifying handwritten digits by comparing them to each other: given an image of an 8, this image should be closer to the average image of an 8 than to any other digit. The following `average_digit` function does this for you.

Listing 5.3   Computing the average value for images representing the same digit

```
import numpy as np
from dlgo.nn.load_mnist import load_data
from dlgo.nn.layers import sigmoid_double

def average_digit(data, digit):
    filtered_data = [x[0] for x in data if np.argmax(x[1]) == digit]
    filtered_array = np.asarray(filtered_data)
    return np.average(filtered_array, axis=0)

train, test = load_data()
avg_eight = average_digit(train, 8)
```

Compute the average over all samples in your data representing a given digit.

Use the average 8 as parameters for a simple model to detect 8s.

What does the average 8 in your training set look like? Figure 5.2 gives the answer.

Figure 5.2   This is what an average handwritten 8 from the MNIST training set looks like. Averaging many hundreds of images in general will result in an unrecognizable blob, but this average 8 still looks very much like an 8.

Because handwriting can differ quite a lot from individual to individual, as expected, the average 8 is a little fuzzy, but it's still noticeably shaped like an 8. Maybe you can use this representation to identify other 8s in your data set? You use the following code to compute and display figure 5.2.

> **Listing 5.4   Computing and displaying the average 8 in your training set**

```
from matplotlib import pyplot as plt

img = (np.reshape(avg_eight, (28, 28)))
plt.imshow(img)
plt.show()
```

This average representation of an 8, `avg_eight`, in the training set of MNIST, should carry a lot of information about what it means for an image to have an 8 on it. You'll use `avg_eight` as *parameters* of a simple model to decide whether a given input vector x, representing a digit, is an 8. In the context of neural networks, we often speak of *weights* when referring to parameters, and `avg_eight` will serve as your weight.

For convenience, you'll use transposition and define `W = np.transpose(avg _eight)`. You can then compute the *dot product* of W and x, which does pointwise multiplication of values of W and x and sums up all 784 resulting values. If your heuristic is right, if x is an 8, individual pixels should have a darker tone at roughly the same places W does, and vice versa. Conversely, if x isn't an 8, there should be less overlap. Let's test this hypothesis in a few examples.

> **Listing 5.5   Computing how close a digit is to your weights by using the dot product**

**Training sample
at index 2 is a 4**

```
x_3 = train[2][0]
x_18 = train[17][0]
```
**Training sample at
index 17 is an 8**

```
W = np.transpose(avg_eight)
np.dot(W, x_3)          ◁──── This evaluates to about 20.1.
np.dot(W, x_18)  ◁─┐
```
**This term is much
bigger, about 54.2.**

You compute the dot product of your weights W with two MNIST samples, one representing a 4, and another one representing an 8. You can see that the latter result of 54.2 for the 8 is much higher than the 20.1 result for the 4. So, it seems you're on to something. Now, how do you decide when a resulting value is high enough to predict it as an 8? In principle, the dot product of two vectors can spit out any real number. What you do to address this is *transform* the output of the dot product to the range [0, 1]. Doing so, you can, for instance, try to define a cutoff value at 0.5 and declare everything above this value an 8.

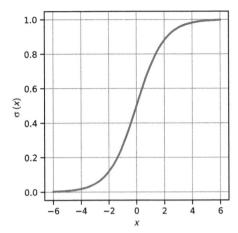

Figure 5.3 Plot of the sigmoid function. The sigmoid maps real values to a range of [0, 1]. Around 0, the slope is rather steep, and the curve flattens out both for small and large values.

One way to do this is with the *sigmoid* function. The sigmoid function is often denoted by σ, the Greek letter *sigma*. For a real number x, the sigmoid function is defined as:

$$\sigma(x) = \frac{1}{1 + e^{-x}}$$

Figure 5.3 shows how it looks to gain some intuition.

Next, let's code the sigmoid function in Python before you apply it to the output of the dot product.

Listing 5.6 Simple implementation of sigmoid function for double values and vectors

```
def sigmoid_double(x):
    return 1.0 / (1.0 + np.exp(-x))


def sigmoid(z):
    return np.vectorize(sigmoid_double)(z)
```

Note that you provide both `sigmoid_double`, which operates on double values, as well as a version that computes the sigmoid for vectors that you'll use extensively in this chapter. Before you apply sigmoid to your previous computation, note that the sigmoid of 2 is already close to 1, so for your two previously computed samples, sigmoid(54.2) and sigmoid(20.1) will be practically indistinguishable. You can fix this issue by *shifting* the output of the dot product toward 0. Doing this is called applying a *bias term*, which we often refer to as b. From the samples, you compute that a good guess for a bias term might be $b = -45$. Using weights and the bias term, you can now compute *predictions* of your model as follows.

Listing 5.7   Computing predictions from weights and bias with dot product and sigmoid

```
def predict(x, W, b):                              ◁        A simple prediction is defined by
    return sigmoid_double(np.dot(W, x) + b)                 applying sigmoid to the output of
                                                            np.doc(W, x) + b.

  b = -45                                The prediction for
                                         the example with
  print(predict(x_3, W, b))    ◁────     a 4 is close to 0.     The prediction for an 8 is
  print(predict(x_18, W, b))   ◁                                0.96 here. You seem to be
                                                                onto something with your
Based on the examples computed                                 heuristic.
so far, you set the bias term to −45.
```

You get satisfying results on the two examples x_3 and x_18. The prediction for the latter is close to 1, and for the former is almost 0. This procedure of mapping an input vector x to $\sigma(Wx + b)$ for W, a vector the same size as x, is called *logistic regression*. Figure 5.4 depicts this algorithm schematically for a vector of length 4.

To get a better idea of how well this procedure works, let's compute predictions for all training and test samples. As indicated before, you define a cutoff, or decision *threshold*, to decide whether a prediction is counted as an 8. As an evaluation metric here, you choose *accuracy*; you compute the ratio of correct predictions among all predictions.

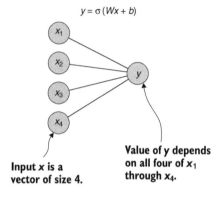

Figure 5.4   An example of logistic regression, mapping an input vector x of length 4 to an output value y between 0 and 1. The schematic indicates how the output y depends on all four values in the input vector x.

Listing 5.8   Evaluating predictions of your model with a decision threshold

```
def evaluate(data, digit, threshold, W, b):        ◁       For an evaluation metric, you
    total_samples = 1.0 * len(data)                        choose accuracy, the ratio of
    correct_predictions = 0                                correct predictions among all.
    for x in data:
        if predict(x[0], W, b) > threshold and np.argmax(x[1]) == digit:
            correct_predictions += 1
        if predict(x[0], W, b) <= threshold and np.argmax(x[1]) != digit:   ◁
            correct_predictions += 1
    return correct_predictions / total_samples      If the prediction is below your
                                                    threshold and the sample isn't
                                                    an 8, you also predicted correctly.
```

Predicting an instance of an 8 as 8 is a correct prediction.

Let's use this evaluation function to assess the quality of predictions for three data sets: the training set, the test set, and the set of all 8s among the test set. You do this with a threshold of 0.5 and weights W and bias term b as before.

Listing 5.9 Calculating prediction accuracy for three data sets

Accuracy on training data of your simple model is 78% (0.7814).

Accuracy on test data is slightly lower, at 77% (0.7749).

```
evaluate(data=train, digit=8, threshold=0.5, W=W, b=b)

evaluate(data=test, digit=8, threshold=0.5, W=W, b=b)

eight_test = [x for x in test if np.argmax(x[1]) == 8]
evaluate(data=eight_test, digit=8, threshold=0.5, W=W, b=b)
```

Evaluating only on the set of 8s in the test set results in only 67% accuracy (0.6663).

What you see is that accuracy on the training set is highest, at about 78%. This shouldn't come as a surprise, because you *calibrated* your model on the training set. In particular, it doesn't make sense to evaluate on the training set, because it doesn't tell you how well your algorithm *generalizes* (how well it performs on previously unseen data). Performance on test data is close to that for training, at about 77%, but it's the last accuracy term that's noteworthy. On the set of all 8s in the test set, you achieve merely 66%, so with your simple model, you're right in only two out of three unseen cases of 8s. This result might be acceptable as a first baseline, but is far from the best you can do. What went wrong, and what can you do better?

- Your model is capable of distinguishing between only a specific digit (here, an 8) and all others. Because in both the training and test sets the number of images per digit is *balanced*, only about 10% are 8s. Thus, a model predicting 0 all the time would yield about 90% accuracy. Keep in mind such *class imbalances* when analyzing classification problems like this one. In light of this, your 77% accuracy on test data doesn't look quite as strong anymore. *You need to define a model that can predict all 10 digits accurately.*

- The parameters of your models are fairly small. For a collection of many thousand diverse handwritten images, all you have is a set of weights the size of one of such image. It's unrealistic to believe you can capture the variability in handwriting found on these images with such a small model. *You have to find a class of algorithms that uses many more parameters effectively to capture the variability in data.*

- For a given prediction, you simply chose a cutoff value to declare a digit an 8 or not. You didn't use the actual prediction value to assess the quality of your model. For instance, a correct prediction at 0.95 certainly indicates a stronger result than one at 0.51. *You have to formalize the notion of how close the prediction was to the actual outcome.*

- You handcrafted the parameters of your model, guided by intuition. Although this might be a good first shot, the promise of machine learning is that you don't have to impose your view on the data, but rather let the algorithm learn from data. Whenever your model makes correct predictions, you need to reinforce this behavior, and whenever the output is wrong, you need to adjust your model accordingly. In other words, *you need to devise a mechanism that updates model parameters according to how well you predicted on training data.*

Although the discussion of this little use case and the naive model you built might not seem like much, you've already seen many parts constituting neural networks. In the next section, you'll use the intuition built around this use case to take your first steps with neural networks by tackling each of these four points.

5.2 The basics of neural networks

How can you improve your OCR model? As we hinted in the introduction, neural networks can do a bang-up job on this sort of task—much better than our handcrafted model. But the handcrafted model does illustrate key concepts that you'll use to build your neural network. This section describes the model from the previous section in the language of neural networks.

5.2.1 Logistic regression as simple artificial neural network

In section 5.1, you saw logistic regression used for *binary classification*. To recap, you took a feature vector x, representing a data sample, fed it into the algorithm by first multiplying it by a weight matrix W and then adding a bias term b. To end up with a prediction y between 0 and 1, you applied the sigmoid function to it: $y = \sigma(Wx + b)$

You should notice a few things here. First of all, the feature vector x can be interpreted as a collection of neurons, sometimes called *units*, connected to y by means of W and b, which you've already seen in figure 5.4. Next, note that the sigmoid can be seen as an activation function, in that it takes the outcome of $Wx + b$ and maps it to the range [0,1]. If you interpret a value close to 1 as the neuron y activating, and in turn not activating if it's close to 0, this setup can be seen as a small example of an artificial neural network already.

5.2.2 Networks with more than one output dimension

In the use case in section 5.1, you simplified the problem of recognizing handwritten digits to a binary classification problem; namely, distinguishing an 8 from all other digits. But you're interested in predicting 10 classes, one for each digit. At least formally, you can achieve this fairly easily by changing what you denote by y, W, and b; you alter the output, weights, and bias of your model.

First, you make y a vector of length 10; y will have one value representing the likelihood of each of the 10 digits:

$$y = \begin{bmatrix} y_0 \\ y_1 \\ \vdots \\ y_8 \\ y_9 \end{bmatrix}$$

Next, let's adapt weights and bias accordingly. Recall that so far W is a vector of length 784. Instead, you make W a matrix with dimensions (10, 784). This way, you can do matrix multiplication of W and an input vector x, namely Wx, the result of which will be a vector of length 10. Continuing, if you make the bias term a vector of length 10, you can add it to Wx. Finally, note that you can compute the sigmoid of a vector z by applying it to each of its components:

$$\sigma(z) = \begin{bmatrix} \sigma(z_0) \\ \sigma(z_1) \\ \vdots \\ \sigma(z_8) \\ \sigma(z_9) \end{bmatrix}$$

Figure 5.5 depicts this slightly altered setup for four input and two output neurons.

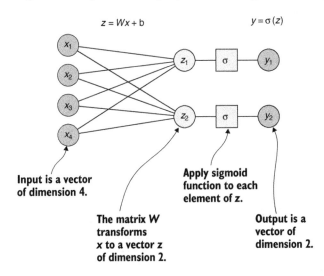

Input is a vector of dimension 4.

The matrix W transforms x to a vector z of dimension 2.

Apply sigmoid function to each element of z.

Output is a vector of dimension 2.

Figure 5.5 In this simple network, four input neurons are connected to two output neurons by means of first multiplying with a two-by-four matrix, adding a two-dimensional bias term, and then applying the sigmoid component-wise.

Now, what did you gain? You can now map an input vector x to an output vector y, whereas before, y was just a single value. The benefit of this is that nothing stops you from doing this vector-to-vector transformation multiple times, thereby building what we call a *feed-forward network.*

5.3 *Feed-forward networks*

Let's quickly recap what you did in the preceding section. On a high level, you carried out the following steps:

1 You started from a vector of input neurons x and applied a simple transformation to it, namely $z = Wx + b$. In the language of linear algebra, these transformations are called *affine linear*. Here you use z as an intermediary variable to ease notation down the line.

2 You applied an activation function, the sigmoid $y = \sigma(z)$, to get output neurons y. The outcome of applying σ tells you how much y activates.

At the heart of feed-forward networks is the idea that you can apply this process iteratively, thereby applying many times over the simple building blocks specified by these two steps. These building blocks form what we call a *layer*. With this notation, you can say that you *stack many layers* to form a *multilayer neural network*. Let's modify our last example by introducing one more layer. You now have to run the following steps:

1 Starting with the input x, compute $z^1 = W^1 x + b^1$.

2 From the intermediate result z^1, you get the output y by computing $y = W^2 z^1 + b^2$.

Note that you use superscripts here to denote which layer you're in, and subscripts to denote position within a vector or matrix. The prescription of working with two layers instead of just one is visualized in figure 5.6.

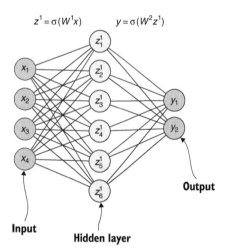

Figure 5.6 An artificial neural network with two layers. The input neurons x connect to an intermediate set of units z, which themselves connect to the output neurons y.

At this point, it should become clear that you're not bound to any particular number of layers to stack. You could use many more. Moreover, you don't necessarily have to use logistic sigmoid as the activation all the time. You have a plethora of activation functions to choose from, and we introduce some of them in the next chapter. Applying these functions of all layers in a network sequentially to one or more data points is usually referred to as a *forward pass*. It's referred to as *forward*, because data always flows only forward, from input to output (in the figures, left to right), and never back.

With this notation, depicting a regular feed-forward network with three layers looks like figure 5.7.

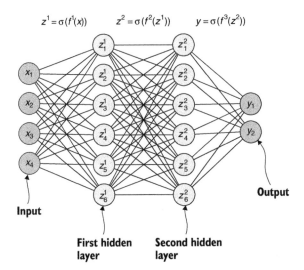

$$z^1 = \sigma(f^1(x)) \qquad z^2 = \sigma(f^2(z^1)) \qquad y = \sigma(f^3(z^2))$$

Input

First hidden layer

Second hidden layer

Output

Figure 5.7 **A neural network with three layers. When defining a neural network, you're limited in neither the number of layers, nor the number of neurons per layer.**

To recap what you've learned so far, let's put together all the notions we mentioned in one concise list:

- A *sequential neural network* is a mechanism to map features, or input neurons, x, to predictions, or output neurons, y. You do this by stacking layers of simple functions one by one in a sequential manner.

- A *layer* is a prescription to map a given input to an output. Computing the output of a layer for a batch of data is called a *forward pass*. Likewise, to compute the forward pass for a sequential network is to compute the forward pass of each layer sequentially, starting with the input.

- The sigmoid function is an activation function that takes a vector of real-valued neurons and *activates* them so that they're mapped to the range $[0,1]$. You interpret values close to 1 as activating.

- Given a weight matrix W and a bias term b, applying the affine-linear transformation $Wx + b$ forms a layer. This kind of layer is usually called a *dense layer* or *fully connected layer*. Going forward, we'll stick to calling them *dense layers*.

- Depending on the implementation, dense layers may or may not come with activations built in; you might see $\sigma(Wx + b)$ as the layer, not just the affine-linear transformation. On the other hand, it's common to consider just the activation function a layer, and you'll do so in your implementation. In the end, whether to add activations into your dense layers or not is just a slightly different view of how to split up and group parts of a function into a logical unit.

- A feed-forward neural network is a sequential network consisting of dense layers with activations. For historical reasons we don't have space to dive into, this architecture is also often called *multilayer perceptron*, or *MLP* for short.

- All neurons that are neither input nor output neurons are called *hidden units*. In contrast, input and output neurons are sometimes called *visible units*. The intuition behind this is that hidden units are *internal to the network*, whereas the visible ones are directly observable. This is somewhat of a stretch, because you normally have access to any part of the system, but it's nevertheless good to know this nomenclature. Consequently, the layers between input and output are called *hidden layers*: each sequential network with at least two layers has at least a hidden one.

- If not stated otherwise, *x* will stand for input to the network, *y* for output of it; sometimes with subscripts to indicate which sample you're considering.

  Stacking many layers to build a large network with many hidden layers is called a *deep neural network*, hence the name *deep learning*.

Nonsequential neural networks

At this point in the book, you're concerned with only *sequential* neural networks, in which the layers form a sequence. In a sequential network, you start with the input, and each following (hidden) layer has precisely one predecessor and one successor, ending in the output layer. This is enough to cover everything you need in order to apply deep learning to the game of Go.

In general, the theory of neural networks allows for arbitrary nonsequential architectures as well. For instance, in some applications it makes sense to concatenate, or add, the output of two layers (you merge two or more previous layers). In such a scenario, you merge multiple inputs and emit one output.

In other applications, it can be useful to split an input into several outputs. In general, a layer can have multiple inputs and outputs. We introduce multi-input and multi-output networks in chapters 11 and 12, respectively.

The setup of a multilayer perceptron with l layers is fully described by the set of weights $W = W^1, ..., W^l$, the set of biases $b = b^1, ..., b^l$, and the set of activation functions chosen for each layer. But a vital ingredient for learning from data and updating parameters is still missing: loss functions and how to optimize them.

5.4   *How good are our predictions? Loss functions and optimization*

Section 5.3 defined how to set up a feed-forward neural network and pass input data through it, but you still don't know how to assess the quality of your predictions. To do so, you need a measure to define how close prediction and actual outcome are.

5.4.1   *What is a loss function?*

To quantify how much you missed your target with your prediction, we introduce the concept of *loss functions*, often called *objective functions*. Let's say you have a feed-forward network with weights W, biases b, and sigmoid activation functions. For a given set of input features X_1, ..., Xk and respective labels $\hat{y}_1$, ..., $\hat{y}_k$ (the symbol $\hat{y}$ for a label is pronounced *y-hat*), using your network you can compute predictions y_1, ..., y_k. In such a scenario, a loss function is defined as follows:

$$\sum_i \text{Loss}(W, b, X_i, \hat{y}_i) = \sum_i \text{Loss}(y_i, \hat{y}_i)$$

Here, $\text{Loss}(y_i, \hat{y}_i) \geq 0$, and Loss is a *differentiable function*. A loss function is a smooth function that assigns a non-negative value to each (prediction, label) pair. The loss of a bunch of features and labels is the sum of losses of the samples. A loss function assesses the fit of your algorithm's parameters, given the data you show it. Your training target is to *minimize loss* by finding good strategies to adapt your parameters.

5.4.2 *Mean squared error*

One example of a widely used loss function is *mean squared error* (*MSE*). Although MSE isn't ideal for our use case, it's one of the most intuitive loss functions to work with. You measure how close your prediction was to the actual label, by measuring the squared distance and averaging over all observed examples. Writing $\hat{y} = \hat{y}_1, ..., \hat{y}_k$ for labels and $y = y_1, ..., y_k$ for predictions, the mean squared error is defined as follows:

$$\text{MSE}(y, \hat{y}) = \frac{1}{2} \sum_{i=1}^{k} (y_i - \hat{y}_i)^2$$

We'll present the benefits and drawbacks of various loss functions after you've seen an application of the theory presented here. For now, let's implement the mean squared error in Python.

Listing 5.10   Mean squared error loss function and its derivative

```
import random
import numpy as np


class MSE:                         ◁── Use the mean squared error
                                        as your loss function.
    def __init__(self):
        pass

    @staticmethod
    def loss_function(predictions, labels):      By defining MSE as 0.5 times
        diff = predictions - labels              the square difference between
        return 0.5 * sum(diff * diff)[0]    ◁─   predictions and labels...

    @staticmethod
    def loss_derivative(predictions, labels):    ...the loss derivative is simply
        return predictions - labels         ◁─   predictions—labels.
```

Note that you implemented not only the loss function itself, but also its derivative with respect to your predictions: `loss_derivative`. This derivative is a vector and is obtained by subtracting labels from predictions.

Next, you'll see how derivatives like this one for MSE play a crucial role in training neural networks.

5.4.3 Finding minima in loss functions

The loss function for a set of predictions and labels gives you information about how well tuned the parameters of your model are. The smaller the loss, the better your predictions, and vice versa. The loss function itself is a function of the parameters of your network. In your MSE implementation, you don't directly supply weights, but they're *implicitly* given through `predictions`, because you use weights to compute them.

In theory, you know from calculus that to minimize the loss, you need to compute its *derivative* and set it to 0. We call the set of parameters at this point a *solution*. Computing the derivative of a function and evaluating it at a specific point is called *computing the gradient*. You've done the first step in computing the derivative in your MSE implementation, but there's more to it. What you aim for is to explicitly compute gradients for all weight and bias terms in your network.

If you need a refresher on the basics of calculus, make sure to check out appendix A. Figure 5.8 shows a surface in three-dimensional space. This surface can be interpreted as a loss function for two-dimensional input. The first two axes represent your weights, and the third, upward-pointing axis indicates the loss value.

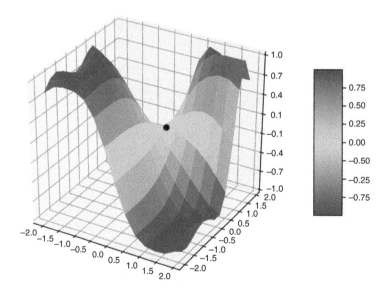

Figure 5.8 An example of a loss function for two-dimensional input (a loss surface). This surface has a minimum around the dark area in the lower right that can be computed by solving the derivative of the loss function.

5.4.4   *Gradient descent to find minima*

Intuitively speaking, when you compute the gradient of a function for a given point, that gradient points in the direction of steepest ascent. Starting with a loss function, Loss, and a set of parameters W, the *gradient descent* algorithm to find a minimum for this function goes like this:

1   Compute the gradient Δ of Loss for the current set of parameters W (compute the derivative of Loss with respect to each weight W).

2   Update W by subtracting Δ from it. We call this step *following the gradient*. Because Δ points in the direction of steepest ascent, subtracting it leads you in the direction of steepest descent.

3   Repeat until Δ is 0.

Because your loss function is non-negative, you know in particular that it has a minimum. It could have many, even infinitely many, minima. For instance, if you think about a flat surface, *every* point on it is a minimum.

Local and global minima

A point with zero gradient reached by gradient descent is by definition a minimum. The precise mathematical definition of a minimum for differentiable functions of many variables is relatively involved and uses information about the *curvature* of the function.

With gradient descent, you'll eventually find a minimum; you can follow the gradient of the function until you find a point of zero gradient. There's only one caveat: you don't know whether this minimum is a *local* or a *global* minimum. You might be stuck in a plateau that locally is the smallest point the function can take, but other points might have a smaller absolute value. The marked point in figure 5.8 is a local minimum, but clearly smaller values exist on this surface.

What we do to resolve this problem might strike you as strange: we ignore it. In practice, gradient descent often leads to satisfying results, so in the context of loss functions for neural networks, we tend to ignore the question of whether a minimum is local or global. We usually won't even run the algorithm until convergence, but rather stop after a predefined number of steps.

Figure 5.9 shows how gradient descent works for the loss surface from figure 5.8 and a choice of parameters indicated by the marked point in the upper right.

In your MSE implementation, you've seen that the derivative of the mean squared error loss is easy to calculate *formally*: it's the difference between labels and predictions. But to *evaluate such a derivative*, you have to compute predictions first. To get a view of the gradients for all parameters, you have to evaluate and aggregate derivatives for every sample in your training set. Given that you're usually dealing with many thousands, if not millions, of data samples for your network, this is practically infeasible. Instead, you approximate gradient computation with a technique called *stochastic gradient descent*.

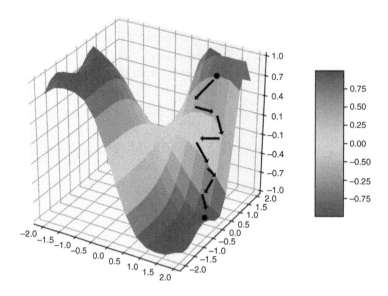

Figure 5.9   Iteratively following the gradients of a loss function will eventually lead you to a minimum.

5.4.5   *Stochastic gradient descent for loss functions*

To compute gradients and apply gradient descent for a neural network, you'd have to evaluate the loss function and its derivative with respect to network parameters at every single point in the training set, which is too expensive in most cases. Instead, you use a technique called *stochastic gradient descent (SGD)*. To run SGD, you first select a few samples from your training set, which you call a *mini-batch*. Each mini-batch is selected with a fixed length that we call the *mini-batch size*. For a classification problem like the handwritten digit problem you're tackling, it's good practice to choose a batch size in the same order of magnitude as the number of labels so as to make sure each label is represented in a mini-batch.

For a given feed-forward neural network with *l* layers and a mini-batch of input data $x_1, \ldots, x_k$ of mini-batch size *k*, you can compute the forward pass of your neural network and compute the loss for that mini-batch. For each sample x_j in this batch, you can then evaluate the gradient of your loss function with respect to any parameter in your network. The weight and bias gradients in layer *i* we call $\Delta_j W^i$ and $\Delta_j b^i$, respectively.

For each layer and each sample in the batch, you compute the respective gradients and use the following *update rules* for parameters:

$$W^i \leftarrow W^i - \alpha \sum_{j=1}^{k} \Delta_j W^i$$

$$b^i \leftarrow b^i - \alpha \sum_{j=1}^{k} \Delta_j b^i$$

You update your parameters by subtracting the cumulative error you receive for that batch. Here $a > 0$ denotes the *learning rate*, an entity specified ahead of training the network.

You'd get much more precise information about gradients if you were to sum over all training samples at once. Using mini-batches is a compromise in terms of gradient accuracy, but is much more computationally efficient. We call this method *stochastic* gradient descent because the mini-batch samples are randomly chosen. Although in gradient descent you have a theoretical guarantee of approaching a local minimum, in SGD that's not the case. Figure 5.10 displays the typical behavior of SGD. Some of your approximate stochastic gradients may not point toward a descending direction, but given enough iterations, you'll usually get close to a (local) minimum.

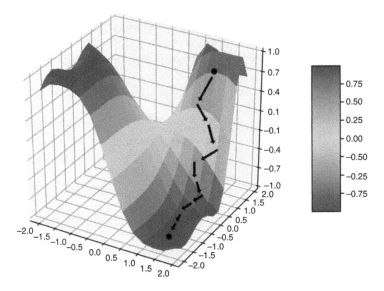

Figure 5.10 Stochastic gradients are less precise, so when following them on a loss surface, you might take a few detours before closing in on a local minimum.

Optimizers

Computing (stochastic) gradients is defined by the fundamental principles of calculus. The way you use the gradients to update parameters isn't. Techniques like the update rule for SGD are called *optimizers*.

Many other optimizers exist, as well as more sophisticated versions of stochastic gradient descent. We cover some of the extensions of SGD in chapter 7. Most of these extensions revolve around adapting the learning rate over time or have more granular updates for individual weights.

5.4.6 *Propagating gradients back through your network*

We discussed how to update parameters of a neural network by using stochastic gradient descent, but we didn't explain how to arrive at the gradients. The algorithm used to compute these gradients is called the *backpropagation* algorithm and is covered in detail in appendix B. This section gives you the intuition behind backpropagation and the necessary building blocks to implement a feed-forward network yourself.

Recall that in a feed-forward network, you run the forward pass on data by computing one simple building block after another. From the output of the last layer, the network's predictions, and the labels, you can compute the loss. The loss function itself is the *composition* of simpler functions. To compute the derivative of the loss function, you can use a fundamental property from calculus: the chain rule. This rule roughly says that the derivative of composed functions is the composition of the derivatives of these functions. Therefore, just as you passed input data forward layer by layer, you can *pass derivatives back layer by layer.* You propagate derivatives back through your network, hence the name *backpropagation.* In figure 5.11, you can see the backpropagation in action for a feed-forward network with two dense layers and sigmoid activations.

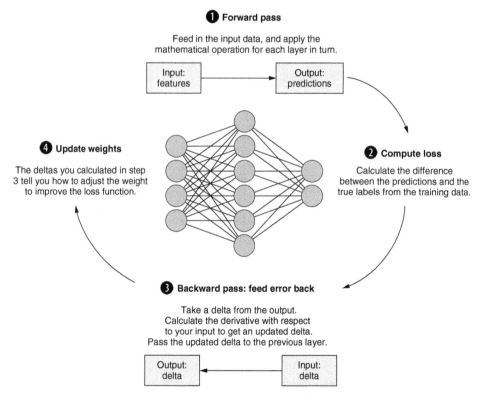

Figure 5.11 Forward and backward passes in a two-layer feed-forward neural network with sigmoid activations and MSE loss function

To guide you through figure 5.11, let's take it step-by-step:

1 *Forward pass on training data.* In this step, you take an input data sample x and pass it through the network to arrive at a prediction, as follows:

   a You compute the affine-linear part: $Wx + b$.

   b You apply the sigmoid function $\sigma(x)$ to the result. Note that we're abusing notation slightly in that x in a computation step means the output of the previous result.

   c You repeat these two steps until you arrive at the output layer. We chose two layers in this example, but the number of layers doesn't matter.

2 *Loss function evaluation.* In this step, you take your labels $\hat{y}$ for the sample x and compare them to your predictions y by computing a loss value. In this example, you choose mean squared error as your loss function.

3 *Propagating error terms back.* In this step, you take your loss value and pass it back through the network. You do so by computing derivatives layer-wise, which is possible because of the chain rule. While the forward pass feeds input data through the network in one direction, the backward pass feeds error terms back in the opposite direction.

   a You propagate error terms, or deltas, denoted by Δ, in the inverse order of the forward pass.

   b To begin with, you compute the derivative of the loss function, which will be your initial Δ. Again, as in the forward pass, we abuse notation and call the propagated error term Δ at every step in the process.

   c You compute the derivative of the sigmoid with respect to its input, which is simply $\sigma \cdot (1 - \sigma)$. To pass Δ to the next layer, you can do component-wise multiplication: $\sigma(1 - \sigma) \cdot \Delta$.

   d The derivative of your affine-linear transformation $Wx + b$ with respect to x is simply W. To pass on Δ, you compute $W^t \cdot \Delta$.

   e These two steps are repeated until you reach the first layer of the network.

4 *Update weights with gradient information.* In the final step, you use the deltas you computed along the way to update your network parameters (weights and bias terms).

   a The sigmoid function doesn't have any parameters, so there's nothing for you to do.

   b The update Δb that the bias term in each layer receives is simply Δ.

   c The update ΔW for the weights in a layer is given by $\Delta \cdot x^T$ (you need to transpose x before multiplying it with delta).

   d Note that we started out by saying x is a single sample. Everything we discussed carries over to mini-batches, however. If x denotes a mini-batch of samples (x is a matrix in which every column is an input vector), the computations of forward and backward passes look exactly the same.

Now that you have all the necessary mathematics covered to build and run a feed-forward network, let's apply what you've learned on a theoretical level by building a neural network implementation from scratch.

5.5   *Training a neural network step-by-step in Python*

The preceding section covered a lot of theoretical ground, but on a conceptual level, you got away with working through only a few basic concepts. For our implementation, you need to worry about only three things: a Layer class, a SequentialNetwork class that's built by adding several Layer objects one by one, and a Loss class that the network needs for backpropagation. These three classes are covered next, after which you'll load and inspect handwritten digit data and apply your network implementation to it. Figure 5.12 shows how those Python classes fit together to implement the forward and backward passes described in the previous section.

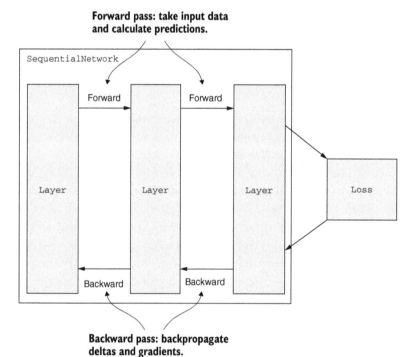

Figure 5.12   Class diagram for your Python implementation of a feed-forward network. A SequentialNetwork contains several Layer instances. Each Layer implements a mathematical function and its derivative. The forward and backward methods implement the forward and backward pass, respectively. A Loss instance calculates your loss function, the error between your prediction and your training data.

5.5.1   *Neural network layers in Python*

To start with a general `Layer` class, note that layers, as we discussed them before, come with not only a prescription to deal with input data (the forward pass), but also a mechanism to *propagate back* error terms. In order not to recompute activation values on the backward pass, it's practical to maintain the *state* of data coming into and out of the layer for both passes. Having said that, the following initialization of `Layer` should be straightforward. You'll begin creating a layers module; later in this chapter you'll use the components in this module to build up a neural network.

Listing 5.11   Base layer implementation

```
import numpy as np

class Layer:
    def __init__(self):
        self.params = []

        self.previous = None
        self.next = None

        self.input_data = None
        self.output_data = None

        self.input_delta = None
        self.output_delta = None
```

Annotations:
- Layers are stacked to build a sequential neural network.
- A layer knows its predecessor (previous)...
- ...and its successor (next).
- Each layer can persist data flowing into and out of it in the forward pass.
- Analogously, a layer holds input and output data for the backward pass.

A layer has a list of parameters and stores both its current input and output data, as well as the respective input and output deltas for the backward pass.

Also, because you're concerned with sequential neural networks, it makes sense to give each layer a successor and a predecessor. Continuing with the definition, you add the following.

Listing 5.12   Connecting layers through successors and predecessors

```
    def connect(self, layer):
        self.previous = layer
        layer.next = self
```

Annotation:
- This method connects a layer to its direct neighbors in the sequential network.

Next, you provide stubs for forward and backward passes in an abstract `Layer` class, which subclasses have to implement.

Listing 5.13   Forward and backward passes in a layer of a sequential neural network

```
    def forward(self):
        raise NotImplementedError

    def get_forward_input(self):
        if self.previous is not None:
```

Annotations:
- Each layer implementation has to provide a function to feed input data forward.
- input_data is reserved for the first layer; all others get their input from the previous output.

```
        return self.previous.output_data
    else:
        return self.input_data

def backward(self):        ◄─────
    raise NotImplementedError
```
Layers have to implement backpropagation of error terms—a way to feed input errors backward through the network.

```
def get_backward_input(self):        ◄───────
    if self.next is not None:
        return self.next.output_delta
    else:
        return self.input_delta
```
Input delta is reserved for the last layer; all other layers get their error terms from their successor.

```
def clear_deltas(self):        ◄──────
    pass
```
You compute and accumulate deltas per mini-batch, after which you need to reset these deltas.

```
def update_params(self, learning_rate):        ◄──┐
    pass
```
Update layer parameters according to current deltas, using the specified learning_rate.

```
def describe(self):
    raise NotImplementedError
```
Layer implementations can print their properties.

As helper functions, you provide get_forward_input and get_backward_input, which just retrieve input for the respective pass, but take special care of input and output neurons. On top of this, you implement a clear_deltas method to reset your deltas periodically, after accumulating deltas over mini-batches, as well as update_params, which takes care of updating parameters for this layer, after the network using this layer tells it to.

Note that as a last piece of functionality, you add a method for a layer to print a description of itself, which you add for convenience to get an easier overview of what your network looks like.

5.5.2   Activation layers in neural networks

Next, you'll provide your first layer, an ActivationLayer. You'll work with the sigmoid function, which you've implemented already. To do backpropagation, you also need its derivative, which can be implemented easily.

> **Listing 5.14   Implementation of the derivative of the sigmoid function**

```
def sigmoid_prime_double(x):
    return sigmoid_double(x) * (1 - sigmoid_double(x))


def sigmoid_prime(z):
    return np.vectorize(sigmoid_prime_double)(z)
```

Note that as for the sigmoid itself, you provide both scalar and vector versions for the derivative. Now, to define an ActivationLayer with the sigmoid function as activation built in, you note that the sigmoid function doesn't have any parameters, so you don't need to worry about updating any parameters just yet.

Listing 5.15   Sigmoid activation layer

```
class ActivationLayer(Layer):
    def __init__(self, input_dim):
        super(ActivationLayer, self).__init__()

        self.input_dim = input_dim
        self.output_dim = input_dim

    def forward(self):
        data = self.get_forward_input()
        self.output_data = sigmoid(data)

    def backward(self):
        delta = self.get_backward_input()
        data = self.get_forward_input()
        self.output_delta = delta * sigmoid_prime(data)

    def describe(self):
        print("|-- " + self.__class__.__name__)
        print("  |-- dimensions: ({},{})"
            .format(self.input_dim, self.output_dim))
```

This activation layer uses the sigmoid function to activate neurons.

The forward pass is simply applying the sigmoid to the input data.

The backward pass is element-wise multiplication of the error term with the sigmoid derivative evaluated at the input to this layer.

Carefully inspect the gradient implementation to see how it fits the picture described in figure 5.11. For this layer, the backward pass is just element-wise multiplication of the layer's current delta with the sigmoid derivative evaluated at the input of this layer: $\sigma(x) \cdot (1 - \sigma(x)) \cdot \Delta$.

5.5.3   *Dense layers in Python as building blocks for feed-forward networks*

To continue with your implementation, you next turn to DenseLayer, which is the more complicated layer to implement, but also the last one you'll tackle in this chapter. Initializing this layer takes a few more variables, because this time you also have to take care of the weight matrix, bias term, and their respective gradients.

Listing 5.16   Dense layer weight initialization

```
class DenseLayer(Layer):

    def __init__(self, input_dim, output_dim):

        super(DenseLayer, self).__init__()

        self.input_dim = input_dim
        self.output_dim = output_dim

        self.weight = np.random.randn(output_dim, input_dim)
        self.bias = np.random.randn(output_dim, 1)
```

Dense layers have input and output dimensions.

Randomly initialize weight matrix and bias vector.

```
self.params = [self.weight, self.bias]
```
The layer parameters consist of weights and bias terms.

```
self.delta_w = np.zeros(self.weight.shape)
self.delta_b = np.zeros(self.bias.shape)
```
Deltas for weights and biases are set to 0.

Note that you initialize W and b randomly. There are many ways to initialize the weights of a neural network. Random initialization is an acceptable baseline, but there are many more sophisticated ways to initialize parameters so that they more accurately reflect the structure of your input data.

Parameter initialization as a starting point for optimization

Initializing parameters is an interesting topic, and we'll discuss a few other initialization techniques in chapter 6.

For now, just keep in mind that initialization will influence your learning behavior. If you think about the loss surface in figure 5.10, initialization of parameters means *choosing a starting point* for optimization; you can easily imagine that different starting points for SGD on the loss surface of figure 5.10 may lead to different results. That makes initialization an important topic in the study of neural networks.

Now, the forward pass for a dense layer is straightforward.

Listing 5.17 Dense layer forward pass

The forward pass of the dense layer is the affine-linear transformation on input data defined by weights and biases.

```
def forward(self):
    data = self.get_forward_input()
    self.output_data = np.dot(self.weight, data) + self.bias
```

As for the backward pass, recall that to compute the delta for this layer, you just need to transpose W and multiply it by the incoming delta: $W^t\Delta$. The gradients for W and b are also easily computed: $\Delta W = \Delta yt$ and $\Delta b = \Delta$, where y denotes the input to this layer (evaluated with the data you're currently using).

Listing 5.18 Dense layer backward pass

```
def backward(self):
    data = self.get_forward_input()
    delta = self.get_backward_input()

    self.delta_b += delta

    self.delta_w += np.dot(delta, data.transpose())

    self.output_delta = np.dot(self.weight.transpose(), delta)
```

For the backward pass, you first get input data and delta.

The current delta is added to the bias delta.

Then you add this term to the weight delta.

The backward pass is completed by passing an output delta to the previous layer.

The update rule for this layer is given by accumulating the deltas, according to the learning rate you specify for your network.

Listing 5.19 Dense layer weight update mechanism

```
def update_params(self, rate):              ◁──┐  Using weight and bias deltas, you
    self.weight -= rate * self.delta_w         │  can update model parameters
    self.bias -= rate * self.delta_b           │  with gradient descent.

def clear_deltas(self):                        ◁──┐  After updating parameters,
    self.delta_w = np.zeros(self.weight.shape)    │  you should reset all deltas.
    self.delta_b = np.zeros(self.bias.shape)

def describe(self):                            ◁──┐  A dense layer can be
    print("|--- " + self.__class__.__name__)      │  described by its input
    print("  |-- dimensions: ({},{})"             │  and output dimensions.
        .format(self.input_dim, self.output_dim))
```

5.5.4 *Sequential neural networks with Python*

Having taken care of layers as building blocks for a network, let's turn to the networks themselves. You initialize a sequential neural network by equipping it with an empty list of layers and let it use MSE as the loss function, unless provided otherwise.

Listing 5.20 Sequential neural network initialization

```
class SequentialNetwork:                ◁──┐  In a sequential neural network,
    def __init__(self, loss=None):         │  you stack layers sequentially.
        print("Initialize Network...")
        self.layers = []
        if loss is None:                   ┌─  If no loss function is
            self.loss = MSE()          ◁───┘  provided, MSE is used.
```

Next, you add functionality to add layers one by one.

Listing 5.21 Adding layers sequentially

```
def add(self, layer):                   ◁──┐  Whenever you add a layer, you connect it to
    self.layers.append(layer)              │  its predecessor and let it describe itself.
    layer.describe()
    if len(self.layers) > 1:
        self.layers[-1].connect(self.layers[-2])
```

At the core of your network implementation is the *train* method. You use mini-batches as input: you shuffle training data and split it into batches of size mini_batch_size. To train your network, you feed it one mini-batch after another. To improve learning, you'll feed the network your training data in batches multiple times. We say that we train it for multiple *epochs*. For each mini-batch, you call the train_batch method. If test_data is provided, you evaluate network performance after each epoch.

Listing 5.22   Train method on a sequential network

```
def train(self, training_data, epochs, mini_batch_size,
        learning_rate, test_data=None):
    n = len(training_data)
    for epoch in range(epochs):                          ◁──┐  To train your network, you pass over data
        random.shuffle(training_data)                       │  for as many times as there are epochs.
        mini_batches = [
            training_data[k:k + mini_batch_size] for
            k in range(0, n, mini_batch_size)
        ]
        for mini_batch in mini_batches:
            self.train_batch(mini_batch, learning_rate)  ◁──┐  For each mini-
        if test_data:                                       │  batch, you train
            n_test = len(test_data)                         │  your network.
            print("Epoch {0}: {1} / {2}"
                .format(epoch, self.evaluate(test_data), n_test))
        else:
            print("Epoch {0} complete".format(epoch))
```

Shuffle training data and create mini-batches.

If you provided test data, you evaluate your network on it after each epoch.

Now, your `train_batch` computes forward and backward passes on this mini-batch and updates parameters afterward.

Listing 5.23   Training a sequential neural network on a batch of data

To train the network on a mini-batch, you compute feed-forward and backward pass...

```
def train_batch(self, mini_batch, learning_rate):
    self.forward_backward(mini_batch)

    self.update(mini_batch, learning_rate)      ◁──┐  ...and then update model
                                                    parameters accordingly.
```

The two steps, `update` and `forward_backward`, are computed as follows.

Listing 5.24   Updating rule and feed-forward and backward passes for your network

```
def update(self, mini_batch, learning_rate):
    learning_rate = learning_rate / len(mini_batch)   ◁──┐  A common technique
    for layer in self.layers:                             │  is to normalize the
        layer.update_params(learning_rate)                │  learning rate by the
    for layer in self.layers:                             │  mini-batch size.
        layer.clear_deltas()

def forward_backward(self, mini_batch):
    for x, y in mini_batch:
        self.layers[0].input_data = x
        for layer in self.layers:                  ◁──┐  For each sample in the mini
            layer.forward()                           │  batch, feed the features
        self.layers[-1].input_delta = \              forward layer by layer.
            self.loss.loss_derivative(self.layers[-1].output_data, y)
        for layer in reversed(self.layers):
            layer.backward()                       ◁──┐  Do layer-by-layer backpropagation
                                                      of error terms.
```

Update parameters for all layers.

Clear all deltas in each layer.

Compute the loss derivative for the output data.

The implementation is straightforward, but there are a few noteworthy points to observe. First, you normalize the learning rate by your mini-batch size in order to keep updates small. Second, before computing the full backward pass by traversing layers in reversed order, you compute the loss derivative of the network output, which serves as the first input delta for the backward pass.

The remaining part of your `SequentialNetwork` implementation concerns model performance and evaluation. To evaluate your network on test data, you need to feed this data forward through your network, and this is precisely what `single_forward` does. The evaluation takes place in `evaluate`, and you return the number of correctly predicted results to assess accuracy.

Listing 5.25   Evaluation

```
def single_forward(self, x):          ◁──┐  Pass a single sample forward
    self.layers[0].input_data = x         │  and return the result.
    for layer in self.layers:
            layer.forward()
    return self.layers[-1].output_data

def evaluate(self, test_data):         ◁──  Compute accuracy on test data.
    test_results = [(
        np.argmax(self.single_forward(x)),
        np.argmax(y)
    ) for (x, y) in test_data]
    return sum(int(x == y) for (x, y) in test_results)
```

5.5.5   *Applying your network handwritten digit classification*

Having implemented a feed-forward network, let's return to our initial use case of predicting handwritten digits for the MNIST data set. After importing the necessary classes that you just built, you load MNIST data, initialize a network, add layers to it, and then train and evaluate the network with your data.

To build a network, keep in mind that your input dimension is 784 and your output dimension is 10, the number of digits. You choose three dense layers with output dimensions 392, 196, and 10, respectively, and add sigmoid activations after each of them. With each new dense layer, you are effectively dividing layer capacity in half. The layer sizes and the number of layers are *hyperparameters* for this network. You've chosen these values to set up a network architecture. We encourage you to experiment with other layer sizes to gain intuition about the learning process of a network in relation to its architecture.

Listing 5.26   Instantiating a neural network

```
from dlgo.nn import load_mnist
from dlgo.nn import network
from dlgo.nn.layers import DenseLayer, ActivationLayer
```

```
training_data, test_data = load_mnist.load_data()    ◁─── Load training and test data.

net = network.SequentialNetwork()    ◁─── Initialize a sequential neural network.

net.add(DenseLayer(784, 392))         ◁─┐ You can then add dense and
net.add(ActivationLayer(392))           │ activation layers one by one.
net.add(DenseLayer(392, 196))
net.add(ActivationLayer(196))
net.add(DenseLayer(196, 10))          ┐ The final layer has size 10, the
net.add(ActivationLayer(10))       ◁─┘ number of classes to predict.
```

You train the network on data by calling `train` with all required parameters. You run training for 10 epochs and set the learning rate to 3.0. As mini-batch size, you choose 10, the number of classes. If you were to shuffle training data near perfectly, in most batches each class would be represented, leading to good stochastic gradients.

Listing 5.27   Running a neural network instance on training data

```
net.train(training_data, epochs=10, mini_batch_size=10,
          learning_rate=3.0, test_data=test_data)    ◁─┐ You can now easily train the
                                                        │ model by specifying train
                                                        │ and test data, the number
                                                        │ of epochs, the mini-batch
                                                        │ size, and the learning rate.
```

Now, run this on the command line:

```
python run_network.py
```

That yields the following prompt:

```
Initialize Network...
|--- DenseLayer
  |-- dimensions: (784,392)
|-- ActivationLayer
  |-- dimensions: (392,192)
|--- DenseLayer
  |-- dimensions: (192,10)
|-- ActivationLayer
  |-- dimensions: (10,10)
Epoch 0: 6628 / 10000
Epoch 1: 7552 / 10000
...
```

The numbers you get for each epoch don't matter here, apart from the fact that the result is highly dependent on the initialization of the weights. But it's noteworthy to observe that you often end up with more than 95% accuracy in less than 10 epochs. This is quite an achievement already, especially given that you did this completely from scratch. In particular, this model performs vastly better than your naive model from the beginning of this chapter. Still, you can do much better.

Note that for the use case you studied, you completely disregard the spatial structure of your input images and treat them as vectors. But it should be clear that the neighborhood of a given pixel is important information that should be used. Ultimately, you want to come back to the game of Go, and you've seen throughout chapters 2 and 3 just how important the neighborhood of (a string of) stones is.

In the next chapter, you'll see how to build a particular kind of neural network that's more suitable for detecting patterns in spatial data, such as images or Go boards. This will bring you much closer to developing a Go bot in chapter 7.

5.6 Summary

- A *sequential neural network* is a simple artificial neural network built from a linear stack of layers. You can apply neural networks to a wide variety of machine-learning problems, including image recognition.
- A *feed-forward network* is a sequential network consisting of dense layers with an activation function.
- *Loss functions* assess the quality of our predictions. *Mean squared error* is one of the most common loss functions used in practice. A loss function gives you a rigorous way to quantify the accuracy of your model.
- *Gradient descent* is an algorithm for minimizing a function. Gradient descent involves following the steepest slope of a function. In machine learning, you use gradient descent to find model weights that give you the smallest loss.
- *Stochastic gradient descent* is a variation on the gradient descent algorithm. In stochastic gradient descent, you compute the gradient on a small subset of your training set called a *mini-batch*, and then update the network weights based on each mini-batch. Stochastic gradient descent is normally much faster than regular gradient descent on large training sets.
- With a sequential neural network, you can use the *backpropagation algorithm* to calculate the gradient efficiently. The combination of backpropagation and mini-batches makes training fast enough to be practical on huge data sets.

Designing a neural
network for Go data

This chapter covers

- Building a deep-learning application that can predict the next Go move from data
- Introducing the Keras deep-learning framework
- Understanding convolutional neural networks
- Building neural networks to analyze spatial Go data

In the preceding chapter, you saw the fundamental principles of neural networks in action and implemented feed-forward networks from scratch. In this chapter, you'll turn your attention back to the game of Go and tackle the problem of how to use deep-learning techniques to predict the next move for any given board situation of a Go game. In particular, you'll generate Go game data with tree-search techniques developed in chapter 4 that you can then use to train a neural network. Figure 6.1 gives an overview of the application you're going to build in this chapter.

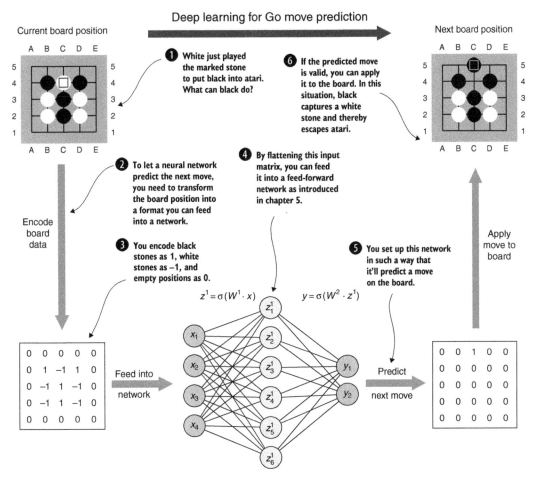

Figure 6.1 How to predict the next move in a game of Go by using deep learning

As figure 6.1 illustrates, to connect your working knowledge of neural networks from the preceding chapter, you have to address a few critical steps first:

1   In chapter 3, you focused on teaching a machine the rules of Go by implementing game play on a Go board. Chapter 4 used these structures for tree search. But in chapter 5, you saw that neural networks need *numerical input;* for the feed-forward architecture you implemented, *vectors* are required.

2   To transform a Go board position into an input vector to be fed into a neural network, you have to create an *encoder* to do the job. In figure 6.1, we sketched a simple encoder that you'll implement in section 6.1; the board is encoded as a matrix of board size, white stones are represented as –1, black stones as 1, and empty points as 0. This matrix can be flattened to a vector, just as you did with MNIST data in the preceding chapter. Although this representation is a little

too simple to provide excellent results for move prediction, it's a first step in the right direction. In chapter 7, you'll see more-sophisticated and useful ways to encode the board.

3 To train a neural network to predict moves, you first have to get your hands on data to feed into it. In section 6.2, you'll pick up the techniques from chapter 4 to generate game records. You'll encode each board position as just discussed, which will serve as your features, and store the next move for each position as labels.

4 Although it's useful to have implemented a neural network as you did in chapter 5, it's now equally important to gain more speed and reliability by introducing a more mature deep-learning library. To this end, section 6.3 introduces *Keras*, a popular deep-learning library written in Python. You'll use Keras to model a network for move prediction.

5 At this point, you might be wondering why you completely discard the spatial structure of the Go board by flattening the encoded board to a vector. In section 6.4, you'll learn about a new layer type called a *convolutional layer* that's much better suited for your use case. You'll use these layers to build a new architecture called a *convolutional neural network*.

6 Toward the end of the chapter, you'll get to know more key concepts of modern deep learning that will further increase move-prediction accuracy, such as efficiently predicting probabilities with *softmax* in section 6.5 or building deeper neural networks in section 6.6 with an interesting activation function called a *rectified linear unit (ReLU)*.

6.1   *Encoding a Go game position for neural networks*

In chapter 3, you built a library of Python classes that represented all the entities in a game of Go: Player, Board, GameState, and so on. Now you want to apply machine learning to problems in Go. But mathematical models like neural networks can't operate on high-level objects like our GameState class; they can deal with only mathematical objects, such as vectors and matrices. In this section, you'll create an Encoder class that translates your native game objects to a mathematical form. Throughout the rest of the chapter, you can feed that mathematical representation to your machine-learning tools.

The first step toward building a deep-learning model for Go move prediction is to load data that can be fed into a neural network. You do this by defining a simple *encoder* for the Go board, introduced in figure 6.1. An encoder is a way to transform the Go board you implemented in chapter 3 in a suitable way. The neural networks you've learned about to this point, multilayer perceptrons, take vectors as inputs, but in section 6.4 you'll see another network architecture that operates on higher-dimensional data. Figure 6.2 gives you an idea how such an encoder could be defined.

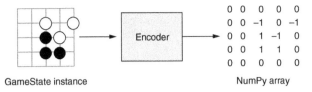

Figure 6.2　An illustration of the `Encoder` class. It takes your `GameState` class and translates it into a mathematical form—a NumPy array.

At its core, an encoder has to know how to encode a full Go game state. In particular, it should define how to encode a single point on the board. Sometimes the inverse is also interesting: if you've predicted the next move with a network, that move will be encoded, and you need to translate it back to an actual move on the board. This operation, called *decoding*, is integral to applying predicted moves.

With this in mind, you can now define your `Encoder` class, an interface for the encoders that you'll create in this and the next chapter. You'll define a new module in dlgo called *encoders*, which you'll initialize with an empty __init__.py, and put the file base.py in it. Then you'll put the following definition in that file.

Listing 6.1　Abstract `Encoder` class to encode Go game state

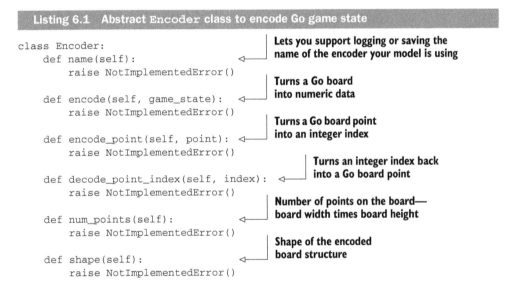

```
class Encoder:
    def name(self):
        raise NotImplementedError()

    def encode(self, game_state):
        raise NotImplementedError()

    def encode_point(self, point):
        raise NotImplementedError()

    def decode_point_index(self, index):
        raise NotImplementedError()

    def num_points(self):
        raise NotImplementedError()

    def shape(self):
        raise NotImplementedError()
```

- Lets you support logging or saving the name of the encoder your model is using
- Turns a Go board into numeric data
- Turns a Go board point into an integer index
- Turns an integer index back into a Go board point
- Number of points on the board— board width times board height
- Shape of the encoded board structure

The definition of encoders is straightforward, but we want to add one more convenience feature into base.py: a function to create an encoder by its name, a string, instead of creating an object explicitly. You do this with the `get_encoder_by_name` function that you append to the definition of encoders.

Listing 6.2   Referencing Go board encoders by name

```
import importlib
```
You can create encoder instances by referencing their name.

```
def get_encoder_by_name(name, board_size):     ◁─┘
    if isinstance(board_size, int):
        board_size = (board_size, board_size)     ◁─┘
    module = importlib.import_module('dlgo.encoders.' + name)
    constructor = getattr(module, 'create')     ◁─┘
    return constructor(board_size)
```
If board_size is one integer, you create a square board from it.

Each encoder implementation will have to provide a "create" function that provides an instance.

Now that you know what an encoder is and how to build one, let's implement the idea from figure 6.2 as your first encoder: one color is represented as 1, the other as −1, and empty points as 0. To make accurate predictions, the model also needs to know whose turn it is. So instead of using 1 for black and −1 for white, you'll use 1 for whoever has the next turn, and −1 for the opponent. You'll call this OnePlaneEncoder, because you encode the Go board into a single matrix or plane of the same size as the board. In chapter 7, you'll see encoders with more *feature planes*; for instance, you'll implement an encoder that has one plane each for black and white stones, and one plane to capture ko. Right now, you'll stick with our simple one-plane encoding idea that you implement in oneplane.py in the encoders module. The following listing shows the first part.

Listing 6.3   Encoding game state with a simple one-plane Go board encoder

```
import numpy as np

from dlgo.encoders.base import Encoder
from dlgo.goboard import Point


class OnePlaneEncoder(Encoder):
    def __init__(self, board_size):
        self.board_width, self.board_height = board_size
        self.num_planes = 1

    def name(self):
        return 'oneplane'

    def encode(self, game_state):     ◁─┘
        board_matrix = np.zeros(self.shape())
        next_player = game_state.next_player
        for r in range(self.board_height):
            for c in range(self.board_width):
                p = Point(row=r + 1, col=c + 1)
                go_string = game_state.board.get_go_string(p)
                if go_string is None:
                    continue
```

To encode, you fill a matrix with 1 if the point contains one of the current player's stones, −1 if the point contains the opponent's stones, and 0 if the point is empty.

You can reference this encoder by the name oneplane.

```
        if go_string.color == next_player:
            board_matrix[0, r, c] = 1
        else:
            board_matrix[0, r, c] = -1
    return board_matrix
```

In the second part of the definition, you'll take care of encoding and decoding single points of the board. The encoding is done by mapping a point on the board to a vector that has a length of board width times board height; the decoding recovers point coordinates from such a vector.

```
    def encode_point(self, point):          ◁— Turns a board point into an integer index
        return self.board_width * (point.row - 1) + (point.col - 1)

    def decode_point_index(self, index):    ◁— Turns an integer index into a board point
        row = index // self.board_width
        col = index % self.board_width
        return Point(row=row + 1, col=col + 1)

    def num_points(self):
        return self.board_width * self.board_height

    def shape(self):
        return self.num_planes, self.board_height, self.board_width
```

This concludes our section on Go board encoders. You can now move on to create data that you can encode and feed into a neural network.

6.2   *Generating tree-search games as network training data*

Before you can apply machine learning to Go games, you need a set of training data. Fortunately, strong players are playing on public Go servers all the time. Chapter 7 covers how to find and process those game records to create training data. For now, you can generate your own game records. This section shows how to use the tree-search bots you created in chapter 4 to generate game records. In the rest of the chapter, you can use those bot game records as training data to experiment with deep learning.

Does it seem silly to use machine learning to imitate a classical algorithm? Not if the traditional algorithm is slow! Here you hope to use machine learning to get a fast approximation to a slow tree search. This concept is a key part of AlphaGo Zero, the strongest version of AlphaGo. Chapter 14 covers how AlphaGo Zero works.

Go ahead and create a file called generate_mcts_games.py outside the dlgo module. As the filename suggests, you'll write code that generates games with MCTS. Each move in each of these games will then be encoded with your `OnePlaneEncoder` from section 6.1 and stored in `numpy` arrays for future use. To begin with, put the following `import` statements at the top of generate_mcts_games.py.

Listing 6.5 Imports for generating encoded Monte Carlo tree-search game data

```
import argparse
import numpy as np

from dlgo.encoders import get_encoder_by_name
from dlgo import goboard_fast as goboard
from dlgo import mcts
from dlgo.utils import print_board, print_move
```

From these imports, you can already see which tools you'll use for the job: the mcts module, your goboard implementation from chapter 3, and the encoders module you just defined. Let's move on to creating the function that'll generate the game data for you. In generate_game, you let an instance of an MCTSAgent from chapter 4 play games against itself (recall from chapter 4 that the *temperature* of an MCTS agent regulates the volatility of your tree search). For each move, you encode the board state before the move has been played, encode the move as a one-hot vector, and then apply the move to the board.

Listing 6.6 Generating MCTS games for this chapter

In boards you store encoded board state; moves is for encoded moves.

```
def generate_game(board_size, rounds, max_moves, temperature):
    boards, moves = [], []

    encoder = get_encoder_by_name('oneplane', board_size)

    game = goboard.GameState.new_game(board_size)

    bot = mcts.MCTSAgent(rounds, temperature)

    num_moves = 0
    while not game.is_over():
        print_board(game.board)
        move = bot.select_move(game)
        if move.is_play:
            boards.append(encoder.encode(game))

            move_one_hot = np.zeros(encoder.num_points())
            move_one_hot[encoder.encode_point(move.point)] = 1
            moves.append(move_one_hot)

        print_move(game.next_player, move)
        game = game.apply_move(move)
        num_moves += 1
        if num_moves > max_moves:
            break

    return np.array(boards), np.array(moves)
```

Initialize a OnePlaneEncoder by name with given board size.

A new game of size board_size is instantiated.

A Monte Carlo tree-search agent with specified number of rounds and temperature will serve as your bot.

The next move is selected by the bot.

The encoded board situation is appended to boards.

The one-hot-encoded next move is appended to moves.

Afterward, the bot move is applied to the board.

You continue with the next move, unless the maximum number of moves has been reached.

Now that you have the means to create and encode game data with Monte Carlo tree search, you can define a main method to run a few games and persist them afterward, which you can also put into generate_mcts_games.py.

Listing 6.7 Main application for generating MCTS games for this chapter

```
def main():
    parser = argparse.ArgumentParser()
    parser.add_argument('--board-size', '-b', type=int, default=9)
    parser.add_argument('--rounds', '-r', type=int, default=1000)
    parser.add_argument('--temperature', '-t', type=float, default=0.8)
    parser.add_argument('--max-moves', '-m', type=int, default=60,
                        help='Max moves per game.')
    parser.add_argument('--num-games', '-n', type=int, default=10)
    parser.add_argument('--board-out')
    parser.add_argument('--move-out')

    args = parser.parse_args()          ◁─────┐ This application allows customization
    xs = []                                   │ via command-line arguments.
    ys = []

    for i in range(args.num_games):
        print('Generating game %d/%d...' % (i + 1, args.num_games))
        x, y = generate_game(args.board_size, args.rounds, args.max_moves,
⮕ args.temperature)              ◁────┐
        xs.append(x)                      │ For the specified number of
        ys.append(y)                      │ games, you generate game data.

    x = np.concatenate(xs)       ◁────┐ After all games have been generated, you
    y = np.concatenate(ys)            │ concatenate features and labels, respectively.

    np.save(args.board_out, x)   ◁────┐ You store feature and label data to separate files,
    np.save(args.move_out, y)         │ as specified by the command-line options.


if __name__ == '__main__':
    main()
```

Using this tool, you can now generate game data easily. Let's say you want to create data for twenty 9 × 9 Go games and store features in features.npy, and labels in labels.npy. The following command will do it:

```
python generate_mcts_games.py -n 20 --board-out features.npy
⮕ --move-out labels.npy
```

Note that generating games like this can be fairly slow, so generating a lot of games will take a while. You could always decrease the number of rounds for MCTS, but this also decreases the bot's level of play. Therefore, we generated game data for you already that you can find in the GitHub repo under generated_games. You can find the output in features-40k.npy and labels-40k.npy; it contains about 40,000 moves over several hundred games. We generated these with 5,000 MCTS rounds per move. At that setting, the MCTS engine mostly plays sensible moves, so we can reasonably hope that a neural network can learn to imitate it.

At this point, you've done all the preprocessing you need in order to apply a neural network to your generated data. You could do this with your network implementation from chapter 5 in a straightforward manner—and it's a good exercise to do so—but going forward, you need a more powerful tool to satisfy your needs to work with increasingly complex deep neural networks. To this end, we introduce Keras next.

6.3   *Using the Keras deep-learning library*

Computing gradients and the backward pass of a neural network is becoming more and more of a lost art form because of the emergence of many powerful deep-learning libraries that hide lower-level abstractions. It's good to have implemented neural networks from scratch in the previous chapter, but now it's time to move on to more mature and feature-rich software.

The Keras deep-learning library is a particularly elegant and popular deep-learning tool written in Python. The open source project was created in 2015 and quickly accumulated a strong user base. The code is hosted at https://github.com/keras-team/keras and has excellent documentation that can be found at https://keras.io.

6.3.1   *Understanding Keras design principles*

One of the strong suits of Keras is that it's an intuitive and easy-to-pick-up API that allows for quick prototyping and a fast experimentation cycle. This makes Keras a popular pick in many data science challenges, such as on https://kaggle.com. Keras is built from modular building blocks and was originally inspired by other deep-learning tools such as Torch. Another big plus for Keras is its extensibility. Adding new custom layers or augmenting existing functionality is relatively straightforward.

Another aspect that makes Keras easy to get started with is that it comes with batteries included. For instance, many popular data sets, like MNIST, can be loaded directly with Keras, and you can find a lot of good examples in the GitHub repository. On top of that, there's a whole community-built ecosystem of Keras extensions and independent projects at https://github.com/fchollet/keras-resources.

A distinctive feature of Keras is the concept of *backends*: it runs with powerful engines that can be swapped on demand. One way to think of Keras is as a deep-learning *front-end*, a library that provides a convenient set of high-level abstractions and functionality to run your models, but is backed by a choice of backend that does the heavy lifting in the background. As of the writing of this book, three official backends are available for Keras: TensorFlow, Theano, and the Microsoft Cognitive Toolkit. In this book, you'll work with Google's TensorFlow library exclusively, which is also the default backend used by Keras. But if you prefer another backend, you shouldn't need much effort to switch; Keras handles most of the differences for you.

In this section, you'll first install Keras. Then you'll learn about its API by running the handwritten digit classification example from chapter 5 with it, and then move on to the task of Go move prediction.

6.3.2 *Installing the Keras deep-learning library*

To get started with Keras, you need to install a backend first. You can start with Tensor-Flow, which is easiest installed through pip by running the following:

```
pip install tensorflow
```

If your machine has an NVIDIA GPU and current CUDA drivers installed, you can try installing the GPU-accelerated version of TensorFlow instead:

```
pip install tensorflow-gpu
```

If `tensorflow-gpu` is compatible with your hardware and drivers, that'll give you a huge speed improvement.

A few optional dependencies that are helpful for model serialization and visualization can be installed for Keras, but you'll skip them for now and directly proceed to installing the library itself:

```
pip install Keras
```

6.3.3 *Running a familiar first example with Keras*

In this section, you'll see that defining and running Keras models follows a four-step workflow:

1. *Data preprocessing*—Load and prepare a data set to be fed into a neural network.
2. *Model definition*—Instantiate a model and add layers to it as needed.
3. *Model compilation*—Compile your previously defined model with an optimizer, a loss function, and an optional list of evaluation metrics.
4. *Model training and evaluation*—Fit your deep-learning model to data and evaluate it.

To get started with Keras, we walk you through an example use case that you encountered in the preceding chapter: predicting handwritten digits with the MNIST data set. As you'll see, our simple model from chapter 5 is remarkably close to the Keras syntax already, so using Keras should come even easier.

With Keras, you can define two types of models: sequential and more general non-sequential models. You'll use only sequential models here. Both model types can be found in keras.models. To define a sequential model, you have to add layers to it, just as you did in chapter 5 in your own implementation. Keras layers are available through the keras.layers module. Loading MNIST with Keras is simple; the data set can be found in the keras.datasets module. Let's import everything you need to tackle this application first.

Listing 6.8   Importing models, layers, and data sets from Keras

```
import keras
from keras.datasets import mnist
from keras.models import Sequential
from keras.layers import Dense
```

Next, you load and preprocess MNIST data, which is achieved in just a few lines. After loading, you flatten the 60,000 training samples and 10,000 test samples, convert them to the float type, and then normalize input data by dividing by 255. This is done because the pixel values of the data set vary from 0 to 255, and you normalize these values to a range of [0, 1], as this will lead to better training of your network. Also, the labels have to be one-hot encoded, just as you did in chapter 5. The following listing shows how to do what we just described with Keras.

Listing 6.9   Loading and preprocessing MNIST data with Keras

```
(x_train, y_train), (x_test, y_test) = mnist.load_data()

x_train = x_train.reshape(60000, 784)
x_test = x_test.reshape(10000, 784)
x_train = x_train.astype('float32')
x_test = x_test.astype('float32')
x_train /= 255
x_test /= 255

y_train = keras.utils.to_categorical(y_train, 10)
y_test = keras.utils.to_categorical(y_test, 10)
```

With data ready to go, you can now proceed to define a neural network to run. In Keras, you initialize a Sequential model and then add layers one by one. In the first layer, you have to provide the input data *shape*, provided through input_shape. In our case, input data is a vector of length 784, so you have to provide input_shape=(784,) as shape information. Dense layers in Keras can be created with an activation keyword to provide the layer with an activation function. You'll choose sigmoid, because it's the only activation function you know so far. Keras has many more activation functions, some of which we'll discuss in more detail.

Listing 6.10   Building a simple sequential model with Keras

```
model = Sequential()
model.add(Dense(392, activation='sigmoid', input_shape=(784,)))
model.add(Dense(196, activation='sigmoid'))
model.add(Dense(10, activation='sigmoid'))
model.summary()
```

The next step in creating a Keras model is to *compile* the model with a loss function and an optimizer. You can do this by specifying strings, and you'll choose sgd (stochastic gradient descent) as the optimizer and mean_squared_error as the loss function. Again, Keras has many more losses and optimizers, but to get started, you'll use the ones you already encountered in chapter 5. Another argument that you can feed into the compilation step of Keras models is a list of evaluation metrics. For your first application, you'll use accuracy as the only metric. The accuracy metric indicates how often the model's highest-scoring prediction matches the true label.

Listing 6.11 Compiling a Keras deep-learning model

```
model.compile(loss='mean_squared_error',
              optimizer='sgd',
              metrics=['accuracy'])
```

The final step for this application is to carry out the training step of the network and then evaluate it on test data. This is done by calling fit on your model by providing not only training data, but also the mini-batch size to work with and the number of epochs to run.

Listing 6.12 Training and evaluating a Keras model

```
model.fit(x_train, y_train,
          batch_size=128,
          epochs=20)
score = model.evaluate(x_test, y_test)
print('Test loss:', score[0])
print('Test accuracy:', score[1])
```

To recap, building and running a Keras model proceeds in four steps: data preprocessing, model definition, model compilation, and model training plus evaluation. One of the core strengths of Keras is that this four-step cycle can be done quickly, which leads to a fast experimentation cycle. This is of great importance, because often your initial model definition can be improved a lot by tweaking parameters.

6.3.4 *Go move prediction with feed-forward neural networks in Keras*

Now that you know what the Keras API for sequential neural networks looks like, let's turn back to our Go move-prediction use case. Figure 6.3 illustrates this step of the process. You'll first load the generated Go data from section 6.2, as shown in listing 6.13. Note that, as with MNIST before, you need to flatten Go board data into vectors.

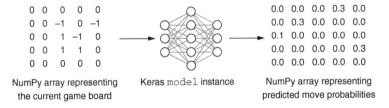

Figure 6.3 A neural network can predict game moves. Having already encoded the game state as a matrix, you can feed that matrix to the move-prediction model. The model outputs a vector representing the probability of each possible move.

Listing 6.13 Loading and preprocessing previously stored Go game data

```
import numpy as np
from keras.models import Sequential
from keras.layers import Dense

np.random.seed(123)
X = np.load('../generated_games/features-40k.npy')
Y = np.load('../generated_games/labels-40k.npy')
samples = X.shape[0]
board_size = 9 * 9

X = X.reshape(samples, board_size)
Y = Y.reshape(samples, board_size)

train_samples = int(0.9 * samples)
X_train, X_test = X[:train_samples], X[train_samples:]
Y_train, Y_test = Y[:train_samples], Y[train_samples:]
```

By setting a random seed, you make sure this script is exactly reproducible.

Load the sample data into NumPy arrays.

Transform the input into vectors of size 81, instead of 9 × 9 matrices.

Hold back 10% of the data for a test set; train on the other 90%.

Next, let's define and run a model to predict Go moves for the features X and labels Y you just defined. For a 9 × 9 board, there are 81 possible moves, so you need to predict 81 classes with your network. As a baseline, pretend you just closed your eyes and pointed at a spot on the board at random. There's a 1 in 81 chance you'd find the next play by pure luck, or 1.2%. So you'd like to see your model significantly exceed 1.2% accuracy.

You define a simple Keras MLP with three Dense layers, each with sigmoid activation functions, that you compile with mean squared error loss and a stochastic gradient descent optimizer. You then let this network train for 15 epochs and evaluate it on test data.

Listing 6.14 Running a Keras multilayer perceptron on generated Go data

```
model = Sequential()
model.add(Dense(1000, activation='sigmoid', input_shape=(board_size,)))
model.add(Dense(500, activation='sigmoid'))
model.add(Dense(board_size, activation='sigmoid'))
model.summary()

model.compile(loss='mean_squared_error',
              optimizer='sgd',
              metrics=['accuracy'])

model.fit(X_train, Y_train,
          batch_size=64,
          epochs=15,
          verbose=1,
          validation_data=(X_test, Y_test))

score = model.evaluate(X_test, Y_test, verbose=0)
print('Test loss:', score[0])
print('Test accuracy:', score[1])
```

Running this code, you should see the model summary and evaluation metrics printed to the console:

Layer (type)	Output Shape	Param #
dense_1 (Dense)	(None, 1000)	82000
dense_2 (Dense)	(None, 500)	500500
dense_3 (Dense)	(None, 81)	40581

```
Total params: 623,081
Trainable params: 623,081
Non-trainable params: 0
```

...

```
Test loss: 0.0129547887068
Test accuracy: 0.0236486486486
```

Note the line `Trainable params: 623,081` in the output; this means the training process is updating the value of over 600,000 individual weights. This is a rough indicator of the computational intensity of the model. It also gives you a rough sense of the *capacity* of your model: its ability to learn complex relationships. As you compare different network architectures, the total number of parameters provides a way to approximately compare the total size of the models.

As you can see, the prediction accuracy of your experiment is at only around 2.3%, which isn't satisfying at first sight. But recall that your baseline of randomly guessing moves is about 1.2%. This tells you that although the performance isn't great, the model is learning and can predict moves better than random.

You can get some insight into the model by feeding it sample board positions. Figure 6.4 shows a board that we contrived to make the right play obvious. Whoever plays next can capture two opponent stones by playing at either A or B. This position doesn't appear in our training set.

Now you can feed that board position into the trained model and print out its predictions.

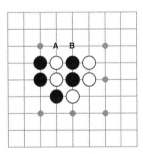

Figure 6.4 An example game position for testing our model. In this position, black can capture two stones by playing at A, or white can capture two stones by playing at B. Whoever plays first in that area has a huge advantage in the game.

Listing 6.15 Evaluating the model on a known board position

```
test_board = np.array([[
    0, 0,  0,   0,   0, 0, 0, 0, 0,
    0, 0,  0,   0,   0, 0, 0, 0, 0,
    0, 0,  0,   0,   0, 0, 0, 0, 0,
    0, 1, -1,   1,  -1, 0, 0, 0, 0,
    0, 1, -1,   1,  -1, 0, 0, 0, 0,
    0, 0,  1,  -1,   0, 0, 0, 0, 0,
    0, 0,  0,   0,   0, 0, 0, 0, 0,
    0, 0,  0,   0,   0, 0, 0, 0, 0,
    0, 0,  0,   0,   0, 0, 0, 0, 0,
]])
move_probs = model.predict(test_board)[0]
i = 0
for row in range(9):
    row_formatted = []
    for col in range(9):
        row_formatted.append('{:.3f}'.format(move_probs[i]))
        i += 1
    print(' '.join(row_formatted))
```

The output looks something like this:

```
0.037 0.037 0.038 0.037 0.040 0.038 0.039 0.038 0.036
0.036 0.040 0.040 0.043 0.043 0.041 0.042 0.039 0.037
0.039 0.042 0.034 0.046 0.042 0.044 0.039 0.041 0.038
0.039 0.041 0.044 0.046 0.046 0.044 0.042 0.041 0.038
0.042 0.044 0.047 0.041 0.045 0.042 0.045 0.042 0.040
0.038 0.042 0.045 0.045 0.045 0.042 0.045 0.041 0.039
0.036 0.040 0.037 0.045 0.042 0.045 0.037 0.040 0.037
0.039 0.040 0.041 0.041 0.043 0.043 0.041 0.038 0.037
0.036 0.037 0.038 0.037 0.040 0.039 0.037 0.039 0.037
```

This matrix maps to the original 9 × 9 board: each number represents the model's confidence that it should play on that point next. This result isn't too impressive; it hasn't even learned not to play on a spot where there's already a stone. But notice that the scores for the edge of the board are consistently lower than scores closer to the center. The conventional wisdom in Go is that you should avoid playing on the very edge of the board, except at the end of the game and other special situations. So the model has learned a legitimate concept about the game: not by understanding strategy or efficiency, but just by copying what our MCTS bot does. This model isn't likely to predict many great moves, but it has learned to avoid a whole class of poor moves.

This is real progress, but you can do better. The rest of this chapter addresses shortcomings of your first experiment and improves Go move-prediction accuracy along the way. You'll take care of the following points:

- The data you're using for this prediction task has been *generated by using tree search*, which has a strong element of randomness. Sometimes MCTS engines generate strange moves, especially when they're either far ahead or far behind in the game. In chapter 7, you'll create a deep-learning model from human game play. Of course, humans are also unpredictable, but they're less likely to play nonsense moves.

- The neural network architecture you used can be vastly improved. Multilayer perceptrons aren't well suited to capture Go board data. You have to flatten the two-dimensional board data to a flat vector, thereby losing all spatial information about the board. In section 6.4, you'll learn about a new type of network that's much better at capturing the Go board structure.

- Throughout all networks so far, you used only the `sigmoid` activation function. In sections 6.5 and 6.6, you'll learn about two new activation functions that often lead to better results.

- Up to this point, you've used MSE only as a loss function, which is intuitive, but not well suited for your use case. In section 6.5, you'll use a loss function that's tailored to classification tasks like ours.

Having addressed most of these points, at the end of this chapter you'll be able to build a neural network that can predict moves better than your first shot. You'll learn key techniques to build a significantly stronger bot in chapter 7.

Keep in mind that, ultimately, you're not interested in predicting moves as accurately as possible, but in creating a bot that can play as well as possible. Even if your deep neural networks may never become extraordinarily good at predicting the next move from historical data, the power of deep learning is that they'll still implicitly pick up the *structure of the game* and play reasonable or even very good moves.

6.4 *Analyzing space with convolutional networks*

In Go, you often see particular local patterns of stones over and over again. Human players learn to recognize dozens of these shapes, and often give them evocative names (like *tiger's mouth, bamboo joint,* or my personal favorite, the *rabbitty six*). To make decisions like a human, our Go AI will also have to recognize many local spatial arrangements. A particular type of neural network called a *convolutional network* is specially designed for detecting spatial relationships like this. Convolutional neural networks, or CNNs, have many applications beyond games: you'll find them applied to images, audio, and even text. This section shows how to build CNNs and apply them to Go game data. First, we introduce the concept of convolution. Next, we show how to build CNNs in Keras. Finally, we show useful ways to process the output of a convolutional layer.

6.4.1 *What convolutions do intuitively*

Convolutional layers and the networks we build from them get their name from a traditional operation from computer vision: *convolutions.* Convolutions are a straightforward way to transform an image or apply a filter, if you will. For two matrices of the same size, a simple convolution is computed by doing the following:

1 Multiplying these two matrices element by element
2 Summing up all the values of the resulting matrix

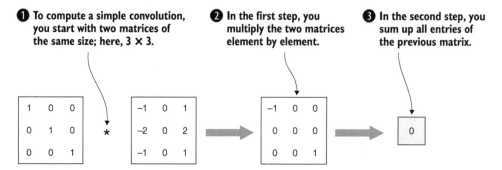

1 To compute a simple convolution, you start with two matrices of the same size; here, 3 × 3.

2 In the first step, you multiply the two matrices element by element.

3 In the second step, you sum up all entries of the previous matrix.

Figure 6.5   In a simple convolution, you multiply two matrices of the same size element by element and then sum up all the values.

The output of such a simple convolution is a scalar value. Figure 6.5 shows an example of such an operation, convolving two 3 × 3 matrices to compute a scalar.

These simple convolutions alone don't help you right away, but they can be used to compute more-complex convolutions that prove useful for your use case. Instead of starting with two matrices of the same size, let's fix the size of the second matrix and increase the size of the first one arbitrarily. In this scenario, you call the first matrix the *input image* and the second one the *convolutional kernel*, or simply *kernel* (sometimes you also see *filter* used). Because the kernel is smaller than the input image, you can compute simple convolutions on many *patches* of the input image. In figure 6.6, you see such a convolution operation of a 10 × 10 input image with a 3 × 3 kernel in action.

The example in figure 6.6 might give you a first hint at why convolutions are interesting for us. The input image is a 10 × 10 matrix consisting of a center 4 × 8 block of 1s surrounded by 0s. The kernel is chosen so that the first column of the matrix (−1, −2, −1) is the negative of the third column (−1, −2, −1), and the middle column is all 0s. Therefore, the following points are true:

- Whenever you apply this kernel to a 3 × 3 patch of the input image in which all pixel values are the same, the output of the convolution will be 0.
- When you apply this convolutional kernel to an image patch in which the left column has higher values than the right, the convolution will be negative.
- When you apply this convolutional kernel to an image patch in which the right column has higher values than the left, the convolution will be positive.

The convolutional kernel is chosen to detect *vertical edges in the input image.* Edges on the left of an object will have positive values; edges on the right will have negative ones. This is exactly what you can see in the result of the convolution in figure 6.6.

The kernel in figure 6.6 is a classical kernel used in many applications and is called a *Sobel kernel.* If you flip this kernel by 90 degrees, you end up with a horizontal edge detector. In the same way, you can define convolutional kernels that blur or sharpen an image, detect corners, and many other things. Many of these kernels can be found in standard image-processing libraries.

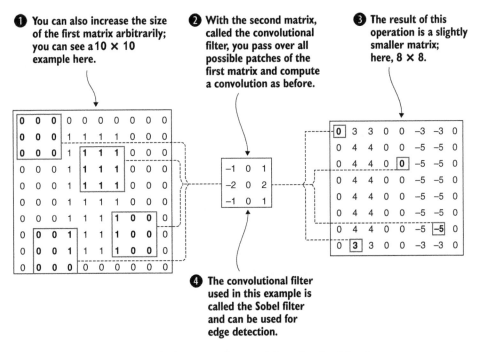

① You can also increase the size of the first matrix arbitrarily; you can see a 10 × 10 example here.

② With the second matrix, called the convolutional filter, you pass over all possible patches of the first matrix and compute a convolution as before.

③ The result of this operation is a slightly smaller matrix; here, 8 × 8.

④ The convolutional filter used in this example is called the Sobel filter and can be used for edge detection.

Figure 6.6 By passing a convolutional kernel over patches of an input image, you can compute a convolution of the image with the kernel. The kernel chosen in this example is a vertical edge detector.

What's interesting is to see that convolutions can be used to extract valuable information from image data, which is exactly what you intend to do for your use case of predicting the next move from Go data. Although in the preceding example we chose a particular convolutional kernel, the way convolutions are used in neural networks is that these kernels are learned from data by backpropagation.

So far, we've discussed how to apply one convolutional kernel to one input image only. In general, it's useful to apply many kernels to many images to produce many output images. How can you do this? Let's say you have four input images and define four kernels. Then you can sum up the convolutions for each input and arrive at one output image. In what follows, you'll call the output images of such convolutions *feature maps*. Now, if you want to have five resulting feature maps instead of one, you define five kernels per input image instead of one. Mapping *n* input images to *m* feature maps, by using *n* × *m* convolutional kernels, is called a *convolutional layer*. Figure 6.7 illustrates this situation.

Seen this way, a convolution layer is a way to transform a number of input images to output images, thereby extracting relevant spatial information of the input. In particular, as you might have anticipated, convolutional layers can be *chained*, thereby forming a neural network of convolutional layers. Usually, a network that consists of convolutional and dense layers only is referred to as *convolutional neural network*, or simply a *convolutional network*.

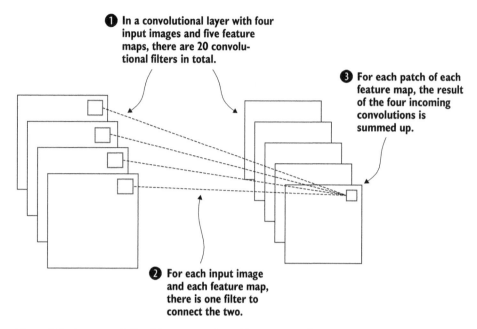

❶ In a convolutional layer with four input images and five feature maps, there are 20 convolutional filters in total.

❸ For each patch of each feature map, the result of the four incoming convolutions is summed up.

❷ For each input image and each feature map, there is one filter to connect the two.

Figure 6.7   In a convolutional layer, a number of input images is operated on by convolutional kernels to produce a specified number of feature maps.

Tensors in deep learning

We stated that the output of a convolutional layer is a bunch of images. Although it can certainly be helpful to think of it that way, there's also a bit more to it. Just as vectors (1D) consist of individual entries, they're not just a bunch of numbers. In the same way, matrices (2D) consist of column vectors, but have an inherent two-dimensional structure that's used in matrix multiplications and other operations (such as convolutions). The output of a convolutional layer has a three-dimensional structure. The filters in a convolutional layer have even one dimension more and possess a 4D structure (a 2D filter for each combination of input and output image). And it doesn't stop there—it's common for advanced deep-learning techniques to deal with even higher-dimensional data structures.

In linear algebra, the higher-dimensional equivalent of vectors and matrices is *tensors*. Appendix A goes into a little more detail, but we can't go into the definition of tensors here. For the rest of this book, you don't need any formal definition of tensors. But apart from having heard about the concept, tensors give us convenient terminology that we use in later chapters. For instance, the collection of images coming out of a convolutional layer can be referred to as 3-Tensor. The 4D filters in a convolutional layer form a 4-Tensor. So you could say that a convolution is an operation in which a 4-Tensor (the convolutional filters) operates on a 3-Tensor (the input images) to transform it into another 3-Tensor.

> **(continued)**
> More generally, you can say that a sequential neural network is a mechanism that transforms tensors of varying dimension step-by-step. This idea of input data "flowing" through a network by using tensors is what led to the name TensorFlow, Google's popular machine-learning library that you'll use to run your Keras models.

Note that in all of this discussion, we've talked only about how to feed data through a convolutional layer, but not how backpropagation would work. We leave this part out on purpose, because it would mathematically go beyond the scope of this book, but more importantly, Keras takes care of the backward pass for us.

Generally, a convolutional layer has a lot fewer parameters than a comparable dense layer. If you were to define a convolutional layer with kernel size $(3, 3)$ on a 28×28 input image, leading to an output of size 26×26, the convolutional layer would have $3 \times 3 = 9$ parameters. In a convolutional layer, you'll usually have a *bias term* as well that's added to the output of each convolution, resulting in a total of 10 parameters. If you compare this to a dense layer connecting an input vector of length 28×28 to an output vector of length 26×26, such a dense layer would have $28 \times 28 \times 26 \times 26 = 529,984$ parameters, excluding biases. At the same time, convolution operations are computationally more costly than regular matrix multiplications used in dense layers.

6.4.2 *Building convolutional neural networks with Keras*

To build and run convolutional neural networks with Keras, you need to work with a new layer type called `Conv2D` that carries out convolutions on two-dimensional data, such as Go board data. You'll also get to know another layer called `Flatten` that flattens the output of a convolutional layer into vectors, which can then be fed into a dense layer.

To start, the preprocessing step for your input data now looks a little different than before. Instead of flattening the Go board, you keep its two-dimensional structure intact.

Listing 6.16 Loading and preprocessing Go data for convolutional neural networks

```
import numpy as np
from keras.models import Sequential
from keras.layers import Dense
from keras.layers import Conv2D, Flatten        Import two new layers, a 2D
                                                convolutional layer, and one that
                                                flattens its input to vectors.

np.random.seed(123)
X = np.load('../generated_games/features-40k.npy')
Y = np.load('../generated_games/labels-40k.npy')

samples = X.shape[0]             The input data shape is three-
size = 9                         dimensional; you use one plane
input_shape = (size, size, 1)    of a 9 × 9 board representation.
```

```
X = X.reshape(samples, size, size, 1)   ◁── Then reshape your input data accordingly.

train_samples = int(0.9 * samples)
X_train, X_test = X[:train_samples], X[train_samples:]
Y_train, Y_test = Y[:train_samples], Y[train_samples:]
```

Now you can use the Keras `Conv2D` object to build the network. You use two convolutional layers, and then *flatten* the output of the second and follow up with two dense layers to arrive at an output of size 9 × 9, as before.

Listing 6.17 Building a simple convolutional neural network for Go data with Keras

The first layer in your network is a Conv2D layer with 48 output filters.

```
  model = Sequential()
└▷ model.add(Conv2D(filters=48,
                    kernel_size=(3, 3),
                    activation='sigmoid',
                    padding='same',
                    input_shape=input_shape))

└▷ model.add(Conv2D(48, (3, 3),
                    padding='same',
                    activation='sigmoid'))

  model.add(Flatten())

  model.add(Dense(512, activation='sigmoid'))
▷ model.add(Dense(size * size, activation='sigmoid'))
  model.summary()
```

For this layer, you choose a 3 × 3 convolutional kernel.

The second layer is another convolution. You leave out the filters and kernel_size arguments for brevity.

Normally, the output of a convolution is smaller than the input. By adding padding='same', you ask Keras to pad your matrix with 0s around the edges, so the output has the same dimension as the input.

You then flatten the 3D output of the previous convolutional layer...

...and follow up with two more dense layers, as you did in the MLP example.

The compiling, running, and evaluating of this model can stay exactly the same as in the MLP example. The only things you changed are the input data shape and the specification of the model itself.

If you run the preceding model, you'll see that the test accuracy has barely budged: it should land somewhere around 2.3% again. That's completely fine—you have a few more tricks to unlock the full power of your convolutional model. For the rest of this chapter, you'll introduce more-advanced deep-learning techniques to improve your move-prediction accuracy.

6.4.3 *Reducing space with pooling layers*

One common technique that you'll find in most deep-learning applications featuring convolutional layers is that of *pooling*. You use pooling to downsize images, to reduce the number of neurons a previous layer has.

The concept of pooling is easily explained: you down-sample images by grouping or pooling patches of the image into a single value. The example in figure 6.8 demonstrates

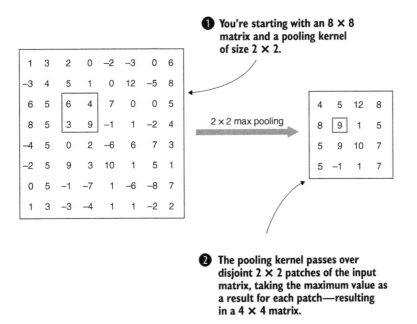

1 You're starting with an 8 × 8 matrix and a pooling kernel of size 2 × 2.

2 x 2 max pooling

2 The pooling kernel passes over disjoint 2 × 2 patches of the input matrix, taking the maximum value as a result for each patch—resulting in a 4 × 4 matrix.

Figure 6.8 Reducing an 8 × 8 image to an image of size (4, 4) by applying a 2 × 2 max pooling kernel

how to cut an image by a factor of 4 by keeping only the maximum value in each disjoint 2 × 2 patch of the image.

This technique is called *max pooling*, and the size of the disjoint patches used for pooling is referred to as *pool size*. You can define other types of pooling as well; for instance, computing the average of the values in a patch. This version is called *average pooling*.

You can define a neural network layer, usually preceding or following a convolutional layer, as follows.

Listing 6.18 Adding a max pooling layer of pool size (2, 2) to a Keras model

```
model.add(MaxPooling2D(pool_size=(2, 2)))
```

You can also experiment with replacing MaxPooling2D by AveragePooling2D in listing 6.4. In cases such as image recognition, pooling is in practice often indispensable to reduce the output size of convolutional layers. Although the operation loses a little information by down-sampling images, it'll usually retain enough of it to make accurate predictions, but at the same time reducing the amount of computation needed quite drastically.

Before you see pooling layers in action, let's discuss a few other tools that will make your Go move predictions much more accurate.

6.5   *Predicting Go move probabilities*

Since we first introduced neural networks in chapter 5, you've used only a single activation function: the logistic sigmoid function. Also, you've been using mean squared error as a loss function throughout. Both choices are good first guesses and certainly have their place in your deep-learning toolbox, but aren't particularly well suited for our use case.

In the end, when predicting Go moves, what you're really after is this question: for each possible move on the board, how *likely* is it that this move is the next move? At each point in time, many good moves are usually available on the board. You set up your deep-learning experiments to find *the* next move from the data you feed into the algorithm, but ultimately the promise of representation learning, and deep learning in particular, is that you can learn enough about the structure of the game to predict the likelihood of a move. You want to predict a *probability distribution* of all possible moves. This can't be guaranteed with sigmoid activation functions. Instead, you introduce the softmax activation function, which is used to predict probabilities in the last layer.

6.5.1   *Using the softmax activation function in the last layer*

The softmax activation function is a straightforward generalization of the logistic sigmoid σ. To compute the softmax function for a vector $x = (x_1, ..., x_l)$, you first apply the exponential function to each component; you compute ex_i. Then you *normalize* each of these values by the sum of all values:

$$\text{softmax}\,(x_i) = \frac{e^{x_i}}{\sum_{j=1}^{l} e^{x_j}}$$

By definition, the components of the softmax function are non-negative and add up to 1, meaning the softmax spits out probabilities. Let's compute an example to see how it works.

Listing 6.19   Defining the softmax activation function in Python

```python
import numpy as np


def softmax(x):
    e_x = np.exp(x)
    e_x_sum = np.sum(e_x)
    return e_x / e_x_sum

x = np.array([100, 100])
print(softmax(x))
```

After defining softmax in Python, you compute it on a vector of length 2; namely, $x = (100, 100)$. If you compute the sigmoid of x, the outcome will be close to $(1, 1)$. But computing the softmax for this example yields $(0.5, 0.5)$. This is what you should've expected: because the values of the softmax function sum up to 1, and both entries are the same, softmax assigns both components equal probability.

Most often, you see the softmax activation function applied as the activation function of the last layer in a neural network, so that you get a guarantee on predicting output probabilities.

> **Listing 6.20 Adding a max pooling layer of pool size (2, 2) to a Keras model**

```
model.add(Dense(9*9, activation='softmax'))
```

6.5.2 *Cross-entropy loss for classification problems*

In the preceding chapter, you started out with mean squared error as your loss function and we remarked that it's not the best choice for your use case. To follow up on this, let's have a closer look at what might go wrong and propose a viable alternative.

Recall that you formulated your move-prediction use case as a *classification problem*, in which you have 9×9 possible classes, only one of which is correct. The correct class is labeled as 1, and all others are labeled as 0. Your predictions for each class will always be a value between 0 and 1. This is a strong assumption on the way your prediction data looks, and the loss function you're using should reflect that. If you look at what MSE does, taking the square of the difference between prediction and label, it makes no use of the fact that you're constrained to a range of 0 to 1. In fact, MSE works best for *regression problems*, in which the output is a continuous range. Think of predicting the height of a person. In such scenarios, MSE will penalize *large* differences. In your scenario, the absolute largest difference between prediction and actual outcome is 1.

Another problem with MSE is that it penalizes all 81 prediction values the same way. In the end, you're concerned only with predicting the one true class, labeled 1. Let's say you have a model that predicts the correct move with a value of 0.6 and all others 0, except for one, which the model assigns to 0.4. In this situation, the mean squared error is $(1 - 0.6)^2 + (0 - 0.4)^2 = 2 \times 0.4^2$, or about 0.32. Your prediction is correct, but you assign the same loss value to both nonzero predictions: about 0.16. Is it really worth putting the same emphasis on the smaller value? If you compare this to the situation in which the correct move gets 0.6 again, but two other moves receive a prediction of 0.2, then the MSE is $(0.4)^2 + 2 \times 0.2^2$, or roughly 0.24, a significantly lower value than in the preceding scenario. But what if the value 0.4 really is more accurate, in that it's just a strong move that *may also be a candidate for the next move?* Should you really penalize this with your loss function?

To take care of these issues, we introduce the *categorical cross-entropy loss function*, or *cross-entropy loss* for short. For labels $\hat{y}$ and predictions y of a model, this loss function is defined as follows:

$$-\sum_i \hat{y}_i \log(y_i)$$

Note that although this might look like a sum consisting of many terms, involving a lot of computation, for our use case this formula boils down to just a single term: the one for which $\hat{y}_i$ is 1. The cross-entropy error is simply $-\log(y_i)$ for the index i for which $\hat{y}_i = 1$. Simple enough, but what do you gain from this?

- Because cross-entropy loss penalizes only the term for which the label is 1, the distribution of all other values isn't directly affected by it. In particular, in the scenario in which you predict the correct next move with a probability of 0.6, there's no difference between attributing one other move a likelihood of 0.4 or two with 0.2. The cross-entropy loss is $-\log(0.6) = 0.51$ in both cases.
- Cross-entropy loss is tailored to a range of [0,1]. If your model predicts a probability of 0 for the move that actually happened, that's as wrong as it can get. You know that $\log(1) = 0$ and that $-\log(x)$ for x between 0 and 1 approaches infinity as x approaches 0, meaning that $-\log(x)$ becomes arbitrarily large (and doesn't just grow quadratically, as MSE).
- On top of that, MSE falls off more quickly as x approaches 1, meaning that you get a much smaller loss for less-confident predictions. Figure 6.9 gives a visual comparison of MSE and cross-entropy loss.

Another crucial point that distinguishes cross-entropy loss from MSE is its behavior during *learning* with stochastic gradient descent (SGD). As a matter of fact, the gradient updates for MSE get smaller and smaller as you approach higher prediction values (y getting closer to 1); learning typically slows down. Compared to this, cross-entropy loss doesn't show this slowdown in SGD, and the parameter updates are proportional to the difference between prediction and true value. We can't go into the details here, but this represents a tremendous benefit for our move-prediction use case.

Compiling a Keras model with categorical cross-entropy loss, instead of MSE, is again simply achieved.

Listing 6.21   Compiling a Keras model with categorical cross-entropy

```
model.compile(loss='categorical_crossentropy'...)
```

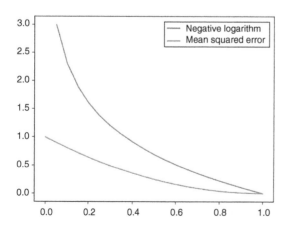

Figure 6.9   Plot of MSE vs. cross-entropy loss for the class labeled as 1. Cross-entropy loss attributes a higher loss for each value in the range [0,1].

With cross-entropy loss and softmax activations in your tool belt, you're now much better equipped to deal with categorical labels and predicting probabilities with a neural network. To finish off this chapter, let's add two techniques that will allow you to build deeper networks—networks with more layers.

6.6 *Building deeper networks with dropout and rectified linear units*

So far, you haven't built a neural network with more than two to four layers. It might be tempting to just add more of the same in the hope that results will improve. It'd be great if it were that simple, but in practice you have a few aspects to consider. Although continually building deeper and deeper neural networks increases the number of parameters a model has and thereby its capacity to adapt to data you feed into it, you may also run into trouble doing so. Among the prime reasons that this might fail is *overfitting*: your model gets better and better at predicting *training* data, but performs suboptimal on *test* data. Put to an extreme, there's no use for a model that can near perfectly predict, or even memorize, what it has seen before, but doesn't know what to do when confronted with data that's a little different. You need to be able to generalize. This is particularly true for predicting the next move in a game as complex as Go. No matter how much time you spend on collecting training data, situations will always arise in game play that your model hasn't encountered before. In any case, it's important to find a strong next move.

6.6.1 *Dropping neurons for regularization*

Preventing overfitting is a common challenge in machine learning in general. You can find a lot of literature about *regularization techniques* that are designed to address the issue of overfitting. For deep neural networks, you can apply a surprisingly simple, yet effective, technique called *dropout*. With dropout applied to a layer in a network, for each training step you pick a number of neurons *at random* and set them to 0; you *drop* these neurons entirely from the training procedure. At each training step, you randomly select new neurons to drop. This is usually done by specifying a *dropout rate*, the percentage of neurons to drop for the layer at hand. Figure 6.10 shows an example dropout layer in which probabilistically half of the neurons are dropped for each mini-batch (forward and backward pass).

The rationale behind this process is that by dropping neurons randomly, you prevent individual layers, and thereby the whole network, from specializing too much to the given data. Layers have to be flexible enough not to rely too much on individual neurons. By doing so, you can keep your neural network from overfitting.

In Keras, you can define a `Dropout` layer with a dropout `rate` as follows.

> **Listing 6.22 Importing and adding a `Dropout` layer to a Keras model**

```
from keras.layers import Dropout

...
model.add(Dropout(rate=0.25))
```

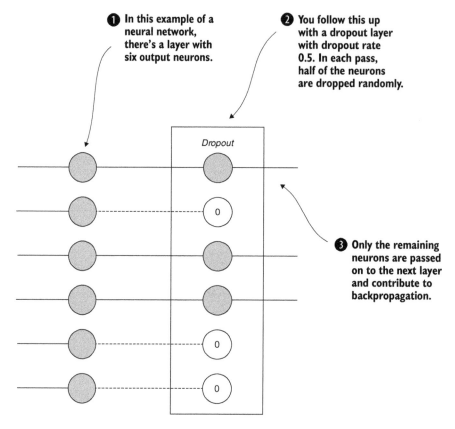

❶ In this example of a neural network, there's a layer with six output neurons.

❷ You follow this up with a dropout layer with dropout rate 0.5. In each pass, half of the neurons are dropped randomly.

Dropout

❸ Only the remaining neurons are passed on to the next layer and contribute to backpropagation.

Figure 6.10   A dropout layer with a rate of 50% will randomly drop half of the neurons from the computation for each mini-batch of data fed into the network.

You can add dropout layers like this in a sequential network before or after every other layer available. Especially in deeper architectures, adding dropout layers is often indispensable.

6.6.2   *The rectified linear unit activation function*

As a last building block for this chapter, you'll get to know the *rectified linear unit* (*ReLU*) activation function, which turns out to often yield better results for deep networks than sigmoid and other activation functions. Figure 6.11 shows what ReLU looks like.

ReLU ignores negative inputs by setting them to 0 and returns positive inputs unchanged. The stronger the positive signal, the stronger the activation with ReLUs. Given this interpretation, rectified linear unit activation functions are pretty close to a simple model of neurons in the brain, in which weaker signals are ignored, but stronger ones lead to the neuron firing. Beyond this basic analogy, we're not going to argue for or against any theoretical benefits of ReLUs, but just note that using them often

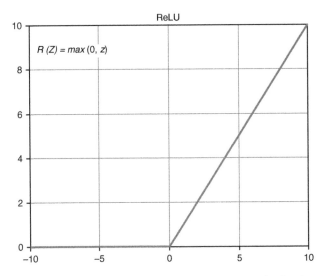

Figure 6.11   The ReLU activation function sets negative inputs to 0 and leaves positive inputs as is.

leads to satisfactory results. To use ReLUs in Keras, you replace `sigmoid` with `relu` in the `activation` argument of your layers.

Listing 6.23   Adding a rectified linear activation to a `Dense` layer

```
from keras.layers import Dense

...
model.add(Dense(activation='relu'))
```

6.7   *Putting it all together for a stronger Go move-prediction network*

The preceding sections covered a lot of ground and introduced not only convolutional networks with max pooling layers, but also cross-entropy loss, the softmax activation for last layers, dropout for regularization, and ReLU activations to improve performance of your networks. To conclude this chapter, let's put every new ingredient you learned about together into a neural network for your Go move-prediction use case and see how well you do now.

To start, let's recall how to load Go data, encoded with your simple one-plane encoder, and reshape it for a convolutional network.

Listing 6.24   Loading and preprocessing Go data for convolutional neural networks

```
import numpy as np
from keras.models import Sequential
from keras.layers import Dense, Dropout, Flatten
from keras.layers import Conv2D, MaxPooling2D
```

```
np.random.seed(123)
X = np.load('../generated_games/features-40k.npy')
Y = np.load('../generated_games/labels-40k.npy')

samples = X.shape[0]
size = 9
input_shape = (size, size, 1)

X = X.reshape(samples, size, size, 1)

train_samples = int(0.9 * samples)
X_train, X_test = X[:train_samples], X[train_samples:]
Y_train, Y_test = Y[:train_samples], Y[train_samples:]
```

Next, let's enhance your previous convolutional network from listing 6.3 as follows:

- Keep the basic architecture intact, starting with two convolutional layers, then a max pooling layer, and two dense layers to finish off.
- Add three dropout layers for regularization: one after each convolutional layer and one after the first dense layer. Use a dropout rate of 50%.
- Change the output layer to a softmax activation, and the inner layers to ReLU activations.
- Change the loss function to cross-entropy loss instead of mean squared error.

Let's see what this model looks like in Keras.

Listing 6.25   Building a convolutional network for Go data with dropout and ReLUs

```
model = Sequential()
model.add(Conv2D(48, kernel_size=(3, 3),
                 activation='relu',
                 padding='same',
                 input_shape=input_shape))
model.add(Dropout(rate=0.5))
model.add(Conv2D(48, (3, 3),
                 padding='same', activation='relu'))
model.add(MaxPooling2D(pool_size=(2, 2)))
model.add(Dropout(rate=0.5))
model.add(Flatten())
model.add(Dense(512, activation='relu'))
model.add(Dropout(rate=0.5))
model.add(Dense(size * size, activation='softmax'))
model.summary()

model.compile(loss='categorical_crossentropy',
              optimizer='sgd',
              metrics=['accuracy'])
```

Finally, to evaluate this model, you can run the following code.

Listing 6.26 Evaluating your enhanced convolutional network

```
model.fit(X_train, Y_train,
          batch_size=64,
          epochs=100,
          verbose=1,
          validation_data=(X_test, Y_test))
score = model.evaluate(X_test, Y_test, verbose=0)
print('Test loss:', score[0])
print('Test accuracy:', score[1])
```

Note that this example increases the number of epochs to 100, whereas you used 15 before. The output looks something like this:

Layer (type)	Output Shape	Param #
conv2d_1 (Conv2D)	(None, 9, 9, 48)	480
dropout_1 (Dropout)	(None, 9, 9, 48)	0
conv2d_2 (Conv2D)	(None, 9, 9, 48)	20784
max_pooling2d_1 (MaxPooling2	(None, 4, 4, 48)	0
dropout_2 (Dropout)	(None, 4, 4, 48)	0
flatten_1 (Flatten)	(None, 768)	0
dense_1 (Dense)	(None, 512)	393728
dropout_3 (Dropout)	(None, 512)	0
dense_2 (Dense)	(None, 81)	41553

```
Total params: 456,545
Trainable params: 456,545
Non-trainable params: 0
```

```
...
Test loss: 3.81980572336
Test accuracy: 0.0834942084942
```

With this model, your test accuracy goes up to over 8%, which is a solid improvement over your baseline model. Also, note the `Trainable params: 456,545` in the output. Recall that your baseline model had over 600,000 trainable parameters. While increasing the accuracy by a factor of three, you also cut the number of weights. This means the credit for the improvement must go to the *structure* of your new model, not just its size.

On the negative side, the training took a lot longer, in large part because you increased the number of epochs. This model is learning more-complicated concepts, and it needs more training passes. If you have the patience to set epochs even higher, you can pick up a few more percentage points of accuracy with this model. Chapter 7 introduces advanced optimizers that can speed up this process.

Next, let's feed the example board to the model and see what moves it recommends:

```
0.000 0.001 0.001 0.002 0.001 0.001 0.000 0.000 0.000
0.001 0.006 0.011 0.023 0.017 0.010 0.005 0.002 0.000
0.001 0.011 0.001 0.052 0.037 0.026 0.001 0.003 0.001
0.002 0.020 0.035 0.045 0.043 0.030 0.014 0.006 0.001
0.003 0.020 0.030 0.031 0.039 0.039 0.018 0.007 0.001
0.001 0.021 0.033 0.048 0.050 0.032 0.017 0.006 0.001
0.001 0.010 0.001 0.039 0.035 0.022 0.001 0.004 0.001
0.000 0.006 0.008 0.017 0.017 0.010 0.007 0.002 0.000
0.000 0.000 0.001 0.001 0.002 0.001 0.001 0.000 0.000
```

The highest-rated move on the board has a score of 0.052—and maps to point A in figure 6.4, where black captures the two white stones. Your model may not be a master tactician yet, but it has definitely learned something about capturing stones! Of course, the results are far from perfect: it still gives high scores to many points that already have stones on them.

At this point, we encourage you to experiment with the model and see what happens. Here are a few ideas to get you started:

- What's most effective on this problem: max pooling, average pooling, or no pooling? (Remember that removing the pooling layer increases the number of trainable parameters in the model; so if you see any extra accuracy, keep in mind that you're paying for it with extra computation.)
- Is it more effective to add a third convolutional layer, or to increase the number of filters on the two layers that are already there?
- How small can you make the second-to-last Dense layer and still get a good result?
- Can you improve the result by changing the dropout rate?
- How accurate can you make the model without using convolutional layers? How does the size and training time of that model compare to your best results with a CNN?

In the next chapter, you'll apply all the techniques you learned here to build a deep-learning Go bot that's trained on *actual game data*, not just simulated games. You'll also see new ways to encode the inputs, which will improve model performance. With these techniques combined, you can build a bot that makes reasonable moves and can at least beat beginner Go players.

6.8 *Summary*

- With encoders, you can transform Go board states into inputs for neural networks, which is an important first step toward applying deep learning to Go.
- Generating Go data with tree search gives you a first Go data set to apply neural networks to.
- Keras is a powerful deep-learning library with which you can create many relevant deep-learning architectures.
- Using convolutional neural networks, you can leverage the spatial structure of input data to extract relevant features.
- With pooling layers, you can reduce image sizes to reduce computational complexity.
- Using softmax activations in the last layer of your network, you can predict output probabilities.
- Working with categorical cross-entropy as a loss function is a more natural choice for Go move-prediction networks than mean squared error. Mean squared error is more useful when you're trying to predict numbers in a continuous range.
- With dropout layers, you have a simple tool to avoid overfitting for deep network architectures.
- Using rectified linear units instead of sigmoid activation functions can bring a significant performance boost.

Learning from data: a deep-learning bot

This chapter covers

- Downloading and processing actual Go game records

- Understanding the standard format for storing Go games

- Training a deep-learning model for move prediction with such data

- Using sophisticated Go board encoders to create strong bots

- Running your own experiments and evaluating them

In the preceding chapter, you saw many essential ingredients for building a deep-learning application, and you built a few neural networks to test the tools you learned about. One of the key things you're still missing is good data to learn from. A supervised deep neural network is only as good as the data you feed it—and so far, you've had only self-generated data at your disposal.

In this chapter, you'll learn about the most common data format for Go data, the Smart Game Format (SGF). You can obtain historical game records in SGF from practically every popular Go server. To power a deep neural network for Go move prediction, in this chapter you'll download many SGF files from a Go server, encode them in a smart way, and train a neural network with this data. The resulting trained network will be much stronger than any previous models in earlier chapters.

Figure 7.1 illustrates what you can build at the end of this chapter.

At the end of this chapter, you can run your own experiments with complex neural networks to build a strong bot completely on your own. To get started, you need access to real-world Go data.

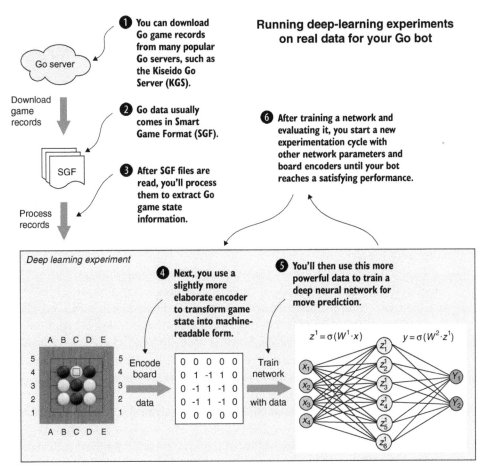

Figure 7.1 Building a deep-learning Go bot, using real-world Go data for training. You can find game records from public Go servers to use for training a bot. In this chapter, you'll learn how to find those records, transform them into a training set, and train a Keras model to imitate the human players' decisions.

7.1   *Importing Go game records*

All the Go data you used up to this point has been generated by yourself. In the preceding chapter, you trained a deep neural network to predict moves for generated data. The best you could hope for was that your network could perfectly predict these moves, in which case the network would play *as well as your tree-search algorithm that generated the data*. In a way, the data you feed into the network provides an upper bound to a deep-learning bot trained from it. The bot can't outperform the strength of the players generating the data.

By using records of games played by strong human opponents as input to your deep neural networks, you can considerably improve the strength of your bots. You'll use game data from the KGS Go Server (formerly known as Kiseido Go Server), one of the most popular Go platforms in the world. Before explaining how to download and process data from KGS, we'll first introduce you to the data format your Go data comes in.

7.1.1   *The SGF file format*

The Smart Game Format (SGF), initially called Smart Go Format, has been developed since the late 80s. Its current, fourth major release (denoted FF[4]) was released in the late 90s. SGF is a straightforward, text-based format that can be used to express games of Go, variations of Go games (for instance, extended game commentaries by professional players), and other board games. For the rest of this chapter, you'll assume that the SGF files you're dealing with consist of Go games without any variations. In this section, we teach you a few basics of this rich game format, but if you want to learn more about it, start with https://senseis.xmp.net/?SmartGameFormat at Sensei's Library.

At its core, SGF consists of metadata about the game and the moves played. You can specify metadata by two capital letters, encoding a property, and a respective value in square brackets. For instance, a Go game played on a board of size (SZ) 9 × 9 would be encoded as SZ[9] in SGF. Go moves are encoded as follows: a white move on the third row and third column of the board is W[cc] in SGF, whereas a black move on the seventh row and third column is represented as B[gc]; the letters B and W stand for stone color, and coordinates for rows and columns are indexed alphabetically. To represent a pass, you use the empty moves B[] and W[].

The following example of an SGF file is taken from the complete 9 × 9 example at the end of chapter 2. It shows a game of Go (Go has game number 1, or GM[1], in SGF) in the current SGF version (FF[4]), played on a 9 × 9 board with zero handicap (HA[0]), and 6.5 points komi for white as compensation for black getting the first move (KM[6.5]). The game is played under Japanese rule settings (RU[Japanese]) and results (RE) in a 9.5-point win by white (RE[W+9.5]):

```
(;FF[4] GM[1] SZ[9] HA[0] KM[6.5] RU[Japanese] RE[W+9.5]
;B[gc];W[cc];B[cg];W[gg];B[hf];W[gf];B[hg];W[hh];B[ge];W[df];B[dg]
;W[eh];B[cf];W[be];B[eg];W[fh];B[de];W[ec];B[fb];W[eb];B[ea];W[da]
```

```
;B[fa];W[cb];B[bf];W[fc];B[gb];W[fe];B[gd];W[ig];B[bd];W[he];B[ff]
;W[fg];B[ef];W[hd];B[fd];W[bi];B[bh];W[bc];B[cd];W[dc];B[ac];W[ab]
;B[ad];W[hc];B[ci];W[ed];B[ee];W[dh];B[ch];W[di];B[hb];W[ib];B[ha]
;W[ic];B[dd];W[ia];B[];
 TW[aa][ba][bb][ca][db][ei][fi][gh][gi][hf][hg][hi][id][ie][if]
  [ih][ii]
 TB[ae][af][ag][ah][ai][be][bg][bi][ce][df][fe][ga]
 W[])
```

An SGF file is organized as a list of *nodes*, which are separated by semicolons. The first node contains metadata about the game: board size, rule set, game result, and other background information. Each following node represents a move in the game. Whitespace is completely unimportant; you could collapse the whole example string into a single line, and it would still be valid SGF. At the end, you also see the points that belong to white's territory, listed under TW, and the ones that belong to black, under TB. Note that the territory indicators are part of the same node as white's last move (W[], indicating a pass): you can consider them as a sort of comment on that position in the game.

This example illustrates some of the core properties of SGF files, and shows everything you'll need for replaying game records in order to generate training data. The SGF format supports many more features, but those are mainly useful for adding commentary and annotations to game records, so you won't need them for this book.

7.1.2 *Downloading and replaying Go game records from KGS*

If you go to the page https://u-go.net/gamerecords/, you'll see a table with game records available for download in various formats (zip, tar.gz). This game data has been collected from the KGS Go Server since 2001 and contains only games played in which at least one of the players was 7 dan or above, or both players were 6 dan. Recall from chapter 2 that dan ranks are master ranks, ranging from 1 dan to 9 dan, so these are games played by strong players. Also note that all of these games were played on a 19 × 19 board, whereas in chapter 6 we used only data generated for the much less complex situation of a 9 × 9 board.

This is an incredibly powerful data set for Go move prediction, which you'll use in this chapter to power a strong deep-learning bot. You want to download this data in an automated fashion by fetching the HTML containing the links to individual files, unpacking the files, and then processing the SGF game records contained in them.

As a first step toward using this data as input to deep-learning models, you create a new submodule called *data* within your main dlgo module, that you provide with an empty __init__.py, as usual. This submodule will contain everything related to Go data processing needed for this book.

Next, to download game data, you create a class called KGSIndex in the new file index_processor.py within the data submodule. Because this step is entirely technical and contributes to neither your Go nor your machine-learning knowledge, we omit the implementation here. If you're interested in the details, the code can be found in our GitHub repository. The KGSIndex implementation found there has precisely one method that you'll use later: download_files. This method will mirror the page https://u-go.net/gamerecords/ locally, find all relevant download links, and then download the respective tar.gz files in a separate folder called data. Here's how you can call it.

Listing 7.1   Creating an index of zip files containing Go data from KGS

```
from dlgo.data.index_processor import KGSIndex


index = KGSIndex()
index.download_files()
```

Running this should result in a command-line output that looks as follows:

```
>>> Downloading index page
KGS-2017_12-19-1488-.tar.gz 1488
KGS-2017_11-19-945-.tar.gz 945


. . .


>>> Downloading data/KGS-2017_12-19-1488-.tar.gz
>>> Downloading data/KGS-2017_11-19-945-.tar.gz


. . .
```

Now that you have this data stored locally, let's move on to processing it for use in a neural network.

7.2   *Preparing Go data for deep learning*

In chapter 6, you saw a simple *encoder* for Go data that was already presented in terms of the Board and GameState classes introduced in chapter 3. When working with SGF files, you first have to extract their content (the *unpacking* we referred to earlier) and replay a game from them, so that you can create the necessary game state information for your Go playing framework.

7.2.1   *Replaying a Go game from an SGF record*

Reading out an SGF file for Go game state information means understanding and implementing the format specifications. Although this isn't particularly hard to do (in the end, it's just imposing a fixed set of rules on a string of text), it's also not the most exciting aspect of building a Go bot and takes a lot of effort and time to do flawlessly. For these reasons, we'll introduce another submodule within dlgo called gosgf that's responsible for handling all the logic of processing SGF files. We treat this submodule as a black box within this chapter and refer you to our GitHub repository for more information on how to read and interpret SGF with Python.

NOTE The gosgf module is adapted from the Gomill Python library, available at https://mjw.woodcraft.me.uk/gomill/.

You'll need precisely one entity from gosgf that's sufficient to process everything you need: Sgf_game. Let's see how you can use Sgf_game to load a sample SGF game, read out game information move by move, and apply the moves to a GameState object. Figure 7.2 shows the beginning of a Go game in terms of SGF commands.

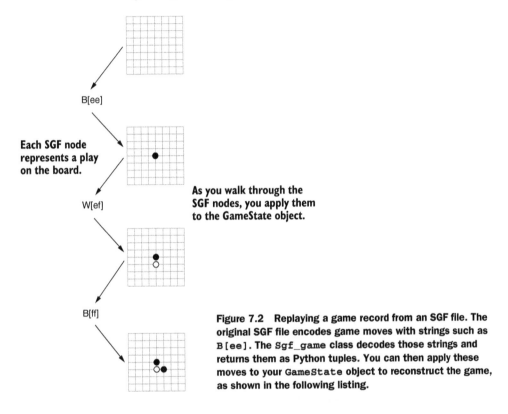

B[ee]

Each SGF node represents a play on the board.

W[ef]

As you walk through the SGF nodes, you apply them to the GameState object.

B[ff]

Figure 7.2 Replaying a game record from an SGF file. The original SGF file encodes game moves with strings such as B[ee]. The Sgf_game class decodes those strings and returns them as Python tuples. You can then apply these moves to your GameState object to reconstruct the game, as shown in the following listing.

Listing 7.2 Replaying moves from an SGF file with your Go framework

```
from dlgo.gosgf import Sgf_game
from dlgo.goboard_fast import GameState, Move
from dlgo.gotypes import Point
from dlgo.utils import print_board
```
◁ **Import the Sgf_game class from the new gosgf module first.**

```
sgf_content = "(;GM[1]FF[4]SZ[9];B[ee];W[ef];B[ff]" + \
              ";W[df];B[fe];W[fc];B[ec];W[gd];B[fb])"
```
◁ **Define a sample SGF string. This content will come from downloaded data later.**

```
sgf_game = Sgf_game.from_string(sgf_content)

game_state = GameState.new_game(19)
```
◁ **With the from_string method, you can create an Sgf_game.**

```
for item in sgf_game.main_sequence_iter():

    color, move_tuple = item.get_move()
    if color is not None and move_tuple is not None:
        row, col = move_tuple
        point = Point(row + 1, col + 1)
        move = Move.play(point)
        game_state = game_state.apply_move(move)
        print_board(game_state.board)
```

Items in this main sequence come as (color, move) pairs, where "move" is a pair of board coordinates.

Iterate over the game's main sequence; you ignore variations and commentaries.

The read-out move can then be applied to your current game state.

In essence, after you have a valid SGF string, you create a game from it, whose main sequence you can iterate over and process however you want. Listing 7.2 is central to this chapter and gives you a rough outline for how you'll proceed to process Go data for deep learning:

1 Download and unpack the compressed Go game files.
2 Iterate over each SGF file contained in these files, read them as Python strings, and create an Sgf_game from these strings.
3 Read out the main sequence of the Go game for each SGF string, make sure to take care of important details such as placing handicap stones, and feed the resulting move data into a GameState object.
4 For each move, encode the current board information as features with an Encoder and store the move itself as a label, before placing it on the board. This way, you'll create move prediction data for deep learning on the fly.
5 Store the resulting features and labels in a suitable format so you can pick it up later and feed it into a deep neural network.

Throughout the next few sections, you'll tackle these five tasks in great detail. After processing data like this, you can go back to your move-prediction application and see how this data affects move-prediction accuracy.

7.2.2   *Building a Go data processor*

In this section, you'll build a Go *data processor* that can transform raw SGF data into features and labels for a machine-learning algorithm. This is going to be a relatively long implementation, so we split it into several parts. When you're finished, you'll have everything ready to run a deep-learning model on real data.

To get started, create a new file called processor.py within your new data submodule. As before, it's also completely fine to just follow the implementation here on a copy of processor.py from the GitHub repository. Let's import a few core Python libraries that you'll work with in processor.py. Apart from NumPy for data, you'll need quite a few packages for processing files.

Listing 7.3   Python libraries needed for data and file processing

```
import os.path
import tarfile
import gzip
import glob
import shutil

import numpy as np
from keras.utils import to_categorical
```

As for functionality needed from dlgo itself, you need to import many of the core abstractions you've built so far.

Listing 7.4   Imports for data processing from the `dlgo` module.

```
from dlgo.gosgf import Sgf_game
from dlgo.goboard_fast import Board, GameState, Move
from dlgo.gotypes import Player, Point
from dlgo.encoders.base import get_encoder_by_name

from dlgo.data.index_processor import KGSIndex
from dlgo.data.sampling import Sampler          ◁─────
```
                                                             Sampler will be used to sample training and test data from files

We haven't yet discussed the two last imports in the listing (Sampler and DataGenerator) but will introduce them as we build our Go data processor. Continuing with processor.py, a GoDataProcessor is initialized by providing an Encoder as string and a data_directory to store SGF data in.

Listing 7.5   Initializing a Go data processor with an encoder and a local data directory

```
class GoDataProcessor:
    def __init__(self, encoder='oneplane', data_directory='data'):
        self.encoder = get_encoder_by_name(encoder, 19)
        self.data_dir = data_directory
```

Next, you'll implement the main data processing method, called load_go_data. In this method, you can specify the number of games you'd like to process, as well as the type of data to load, meaning either *training* or *test* data. load_go_data will download online Go records form KGS, sample the specified number of games, process them by creating features and labels, and then persist the result locally as NumPy arrays.

Listing 7.6   `load_go_data` loads, processes, and stores data

```
    def load_go_data(self, data_type='train',
                     num_samples=1000):           ◁─────
        index = KGSIndex(data_directory=self.data_dir)
        index.download_files()  ◁─┐
```

num_samples refers to the number of games to load data from.

For data_type, you can choose either train or test.

Download all games from KGS to your local data directory. If data is available, it won't be downloaded again.

Collect all zip file names contained in the data in a list.

Group all SGF file indices by zip file name.

```
sampler = Sampler(data_dir=self.data_dir)
data = sampler.draw_data(data_type, num_samples)

zip_names = set()
indices_by_zip_name = {}
for filename, index in data:
    zip_names.add(filename)
    if filename not in indices_by_zip_name:
        indices_by_zip_name[filename] = []
    indices_by_zip_name[filename].append(index)
for zip_name in zip_names:
    base_name = zip_name.replace('.tar.gz', '')
    data_file_name = base_name + data_type
    if not os.path.isfile(self.data_dir + '/' + data_file_name):
        self.process_zip(zip_name, data_file_name,
                         indices_by_zip_name[zip_name])

features_and_labels = self.consolidate_games(data_type, data)
return features_and_labels
```

The Sampler instance selects the specified number of games for a data type.

The zip files are then processed individually.

Features and labels from each zip are then aggregated and returned.

Note that after downloading data, you split it by using a `Sampler` instance. All this sampler does is make sure it randomly picks the specified number of games, but more importantly, that *training and test data don't overlap in any way.* `Sampler` does that by splitting training and test data on a file level, by simply declaring games played prior to 2014 as test data and newer games as training data. Doing so, you make absolutely sure that no game information available in test data is also (partly) included in training data, which may lead to overfitting of your models.

Splitting training and test data

The reason you split data into training and test data is to obtain reliable performance metrics. You train a model on training data and evaluate it on test data to see how well the model adapts to *previously unseen situations,* how well it extrapolates from what it learned in the training phase to the real world. Proper data collection and split is crucially important to trust the results you get from a model.

It can be tempting to just load all the data you have, shuffle it, and randomly split it into training and test data. Depending on the problem at hand, this naive approach may or may not be a good idea. If you think of Go game records, the moves within a single game depend on each other. Training a model on a set of moves that are also included in the test set can lead to the illusion of having found a strong model. But it may turn out that your bot won't be as strong in practice. Make sure to spend time analyzing your data and find a split that makes sense.

After downloading and sampling data, `load_go_data` relies essentially on helpers to process data: `process_zip` to read out individual zip files, and `consolidate_games` to group the results from each zip into one set of features and labels. Let's have a look at `process_zip` next, which carries out the following steps for you:

1. Unzip the current file by using `unzip_data`.
2. Initialize an `Encoder` instance to encode SGF records.
3. Initialize feature and label NumPy arrays of the right shape.
4. Iterate through the game list and process games one by one.
5. For each game, first apply all handicap stones.
6. Then read out each move as found in the SGF record.
7. For each next move, encode the move as `label`.
8. Encode the current board state as `feature`.
9. Apply the next move to the board and proceed.
10. Store small chunks of features and labels in the local filesystem.

Here's how you implement the first nine of these steps in `process_zip`. Note that the technical utility method `unzip_data` has been omitted for brevity, but can be found in our GitHub repository. In figure 7.3, you see how processing zipped SGF files into an encoded game state works.

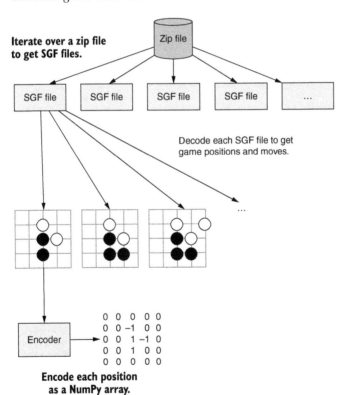

Figure 7.3 The `process_zip` function. You iterate over a zip file that contains many SGF files. Each SGF file contains a sequence of game moves; you use those to reconstruct `GameState` objects. Then you use an `Encoder` object to convert each game state to a NumPy array.

Next, you can define `process_zip`.

Listing 7.7   Processing Go records stored in zip files into encoded features and labels

```
                                                            Determines the total number of
                                                            moves in all games in this zip file
        def process_zip(self, zip_file_name, data_file_name, game_list):
            tar_file = self.unzip_data(zip_file_name)
            zip_file = tarfile.open(self.data_dir + '/' + tar_file)
            name_list = zip_file.getnames()
            total_examples = self.num_total_examples(zip_file, game_list,
                                          name_list)
            shape = self.encoder.shape()
            feature_shape = np.insert(shape, 0, np.asarray([total_examples]))
            features = np.zeros(feature_shape)
            labels = np.zeros((total_examples,))

            counter = 0
            for index in game_list:
                name = name_list[index + 1]
                if not name.endswith('.sgf'):
                    raise ValueError(name + ' is not a valid sgf')
                sgf_content = zip_file.extractfile(name).read()
                sgf = Sgf_game.from_string(sgf_content)

                game_state, first_move_done = self.get_handicap(sgf)

                for item in sgf.main_sequence_iter():
                    color, move_tuple = item.get_move()
                    point = None
                    if color is not None:
                        if move_tuple is not None:
                            row, col = move_tuple
                            point = Point(row + 1, col + 1)
                            move = Move.play(point)
                        else:
                            move = Move.pass_turn()
                        if first_move_done and point is not None:
                            features[counter] = self.encoder.encode(game_state)
                            labels[counter] = self.encoder.encode_point(point)
                            counter += 1
                    game_state = game_state.apply_move(move)
                    first_move_done = True
```

Infers the shape of features and labels from the encoder you use

Reads the SGF content as string, after extracting the zip file

Infers the initial game state by applying all handicap stones

Iterates over all moves in the SGF file

Reads the coordinates of the stone to be played...

Encodes the current game state as features...

...or passes, if there is none

...and the next move as label for the features

Afterward, the move is applied to the board, and you proceed with the next one.

Note how closely the `for` loop resembles the process you sketched in listing 7.2, so this code should feel familiar to you. `process_zip` uses two helper methods that you'll implement next. The first one is `num_total_examples`, which precomputes the number of moves available per zip file so that you can efficiently determine the size of feature and label arrays.

Listing 7.8 Calculating the total number of moves available in the current zip file

```
def num_total_examples(self, zip_file, game_list, name_list):
    total_examples = 0
    for index in game_list:
        name = name_list[index + 1]
        if name.endswith('.sgf'):
            sgf_content = zip_file.extractfile(name).read()
            sgf = Sgf_game.from_string(sgf_content)
            game_state, first_move_done = self.get_handicap(sgf)

            num_moves = 0
            for item in sgf.main_sequence_iter():
                color, move = item.get_move()
                if color is not None:
                    if first_move_done:
                        num_moves += 1
                    first_move_done = True
            total_examples = total_examples + num_moves
        else:
            raise ValueError(name + ' is not a valid sgf')
    return total_examples
```

You use the second helper method to figure out the number of handicap stones the current game has and apply these moves to an empty board.

Listing 7.9 Retrieving handicap stones and applying them to an empty Go board

```
@staticmethod
def get_handicap(sgf):
    go_board = Board(19, 19)
    first_move_done = False
    move = None
    game_state = GameState.new_game(19)
    if sgf.get_handicap() is not None and sgf.get_handicap() != 0:
        for setup in sgf.get_root().get_setup_stones():
            for move in setup:
                row, col = move
                go_board.place_stone(Player.black,
                                     Point(row + 1, col + 1))
        first_move_done = True
        game_state = GameState(go_board, Player.white, None, move)
    return game_state, first_move_done
```

To finish the implementation of process_zip, you store chunks of features and labels in separate files.

Listing 7.10 Persisting features and labels locally in small chunks

```
    feature_file_base = self.data_dir + '/' + data_file_name +
'_features_%d'
    label_file_base = self.data_dir + '/' + data_file_name + '_labels_%d'
```

The current chunk is cut off from features and labels...

You process features and labels in chunks of size 1024.

... and then stored in a separate file.

```
chunk = 0  # Due to files with large content, split up after chunksize
chunksize = 1024
while features.shape[0] >= chunksize:
    feature_file = feature_file_base % chunk
    label_file = label_file_base % chunk
    chunk += 1
    current_features, features = features[:chunksize],
features[chunksize:]
    current_labels, labels = labels[:chunksize], labels[chunksize:]
    np.save(feature_file, current_features)
    np.save(label_file, current_labels)
```

The reason you store small chunks is that NumPy arrays can become large quickly, and storing data in smaller files enables more flexibility later. For instance, you could either consolidate data for all chunks or load chunks into memory as needed. You'll work with both approaches. Although the latter—dynamically loading batches of data as you go— is a little more intricate, consolidating data is straightforward. As a side note, in our implementation you potentially lose the last fraction of a chunk in the `while` loop, but this is insubstantial because you have more than enough data at your disposal.

Continuing with processor.py and our definition of `GoDataProcessor`, you simply *concatenate* all arrays into one.

> **Listing 7.11   Consolidating individual NumPy arrays of features and labels into one set**

```
def consolidate_games(self, data_type, samples):
    files_needed = set(file_name for file_name, index in samples)
    file_names = []
    for zip_file_name in files_needed:
        file_name = zip_file_name.replace('.tar.gz', '') + data_type
        file_names.append(file_name)

    feature_list = []
    label_list = []
    for file_name in file_names:
        file_prefix = file_name.replace('.tar.gz', '')
        base = self.data_dir + '/' + file_prefix + '_features_*.npy'
        for feature_file in glob.glob(base):
            label_file = feature_file.replace('features', 'labels')
            x = np.load(feature_file)
            y = np.load(label_file)
            x = x.astype('float32')
            y = to_categorical(y.astype(int), 19 * 19)
            feature_list.append(x)
            label_list.append(y)
    features = np.concatenate(feature_list, axis=0)
    labels = np.concatenate(label_list, axis=0)
    np.save('{}/features_{}.npy'.format(self.data_dir, data_type),
features)
    np.save('{}/labels_{}.npy'.format(self.data_dir, data_type), labels)

    return features, labels
```

You can test this implementation by loading features and labels for 100 games as follows.

Listing 7.12 Loading training data from 100 game records

```
from dlgo.data.processor import GoDataProcessor

processor = GoDataProcessor()
features, labels = processor.load_go_data('train', 100)
```

These features and labels have been encoded with your oneplane encoder from chapter 6, meaning they have exactly the same structure. In particular, you can go ahead and train any of the networks you created in chapter 6 with the data you just created. Don't expect too much in terms of evaluation performance if you do so. Although this real-world game data is much better than the games generated in chapter 6, you're now working with 19 × 19 Go data, which is much more complex than games played on 9 × 9 boards.

The procedure of loading a lot of smaller files into memory for consolidation can potentially lead to out-of-memory exceptions when loading large amounts of data. You'll address this issue in the next section by using a *data generator* to provide just the next mini-batch of data needed for model training.

7.2.3 *Building a Go data generator to load data efficiently*

The KGS index you downloaded from https://u-go.net/gamerecords/ contains well over 170,000 games, translating into many millions of Go moves to predict. Loading all of these data points into a single pair of NumPy arrays will become increasingly difficult as you load more and more game records. Your approach to consolidate games is doomed to break down at some point.

Instead, we suggest a smart replacement for consolidate_games in your GoData-Processor. Note that in the end, all a neural network needs for training is that you feed it mini-batches of features and labels *one by one*. There's no need to keep data in memory at all times. So, what you're going to build next is a *generator* for Go data. If you know the concept of generators from Python, you'll immediately recognize the pattern of what you're building. If not, think of a generator as a function that efficiently provides you with just the next batch of data you need, when you need it.

To start, let's initialize a DataGenerator. Put this code into generator.py inside the data module. You initialize such a generator by providing a local data_directory and samples as provided by your Sampler in GoDataProcessor.

Listing 7.13 The signature of a Go data generator

```
import glob
import numpy as np
from keras.utils import to_categorical
```

```
class DataGenerator:
    def __init__(self, data_directory, samples):
        self.data_directory = data_directory
        self.samples = samples
        self.files = set(file_name for file_name, index in samples)      <──
        self.num_samples = None

    def get_num_samples(self, batch_size=128, num_classes=19 * 19):      <──┐
        if self.num_samples is not None:
            return self.num_samples
        else:
            self.num_samples = 0
            for X, y in self._generate(batch_size=batch_size,
                                       num_classes=num_classes):
                self.num_samples += X.shape[0]
            return self.num_samples
```

> **Your generator has access to a set of files that you sampled earlier.**

> **Depending on the application, you may need to know how many examples you have.**

Next, you'll implement a private _generate method that creates and returns batches of data. This method follows a similar overall logic as consolidate_games, with one important difference. Whereas previously you created a big NumPy array for both features and labels, you now only return, or yield, the next batch of data.

Listing 7.14   Private method to generate and yield the next batch of Go data

```
    def _generate(self, batch_size, num_classes):
        for zip_file_name in self.files:
            file_name = zip_file_name.replace('.tar.gz', '') + 'train'
            base = self.data_directory + '/' + file_name + '_features_*.npy'
            for feature_file in glob.glob(base):
                label_file = feature_file.replace('features', 'labels')
                x = np.load(feature_file)
                y = np.load(label_file)
                x = x.astype('float32')
                y = to_categorical(y.astype(int), num_classes)
                while x.shape[0] >= batch_size:
                    x_batch, x = x[:batch_size], x[batch_size:]
                    y_batch, y = y[:batch_size], y[batch_size:]
                    yield x_batch, y_batch
```

> **You return, or yield, batches of data as you go.**

All that's missing from your generator is a method to return a generator. Having a generator, you can explicitly call next() on it to generate batches of data for your use case. This is done as follows.

Listing 7.15   Calling the `generate` method to obtain a generator for model training

```
    def generate(self, batch_size=128, num_classes=19 * 19):
        while True:
            for item in self._generate(batch_size, num_classes):
                yield item
```

Before we can show you how to use such a generator to train a neural network, we have to explain how to incorporate this concept into your GoDataProcessor.

7.2.4 *Parallel Go data processing and generators*

You may have noticed that loading just 100 game records in listing 7.3 feels a little slower than you may have expected. Although naturally you need to download the data first, it's the processing itself that's relatively slow. Recall from your implementation that you process zip files *sequentially*. After you finish a file, you proceed to the next. But if you look closely, the processing of Go data as we presented it is what you'd call *embarrassingly parallel*. It takes just a little effort to process the zip files in parallel by distributing workload across all CPUs in your computer; for instance, using Python's multiprocessing library.

In our GitHub repository, you'll find a parallel implementation of GoDataProcessor in the data module in parallel_processor.py. If you're interested in how this works in detail, we encourage you to go through the implementation provided there. The reason we omit the details here is that although the speedup of parallelization is of immediate benefit to you, the implementation details make the code quite a bit harder to read.

Another benefit that you get from using the parallel version of GoDataProcessor is that you can optionally use your DataGenerator with it, to return a generator instead of data.

Listing 7.16 The parallel version of `load_go_data` can optionally return a generator

```
        def load_go_data(self, data_type='train', num_samples=1000,
                         use_generator=False):
            index = KGSIndex(data_directory=self.data_dir)
            index.download_files()

            sampler = Sampler(data_dir=self.data_dir)
            data = sampler.draw_data(data_type, num_samples)

            self.map_to_workers(data_type, data)
            if use_generator:
                generator = DataGenerator(self.data_dir, data)
                return generator
            else:
                features_and_labels = self.consolidate_games(data_type, data)
                return features_and_labels
```

Map workload to CPUs.

Either returns a Go data generator...

...or returns consolidated data as before

With the exception of the use_generator flag in the parallel extension, both GoData-Processor versions share the same interface. Through GoDataProcessor from dlgo .data.parallel_processor, you can now use a generator to provide Go data as follows.

Listing 7.17 Loading training data from 100 game records

```
from dlgo.data.parallel_processor import GoDataProcessor

processor = GoDataProcessor()
generator = processor.load_go_data('train', 100, use_generator=True)
```

```
print(generator.get_num_samples())
generator = generator.generate(batch_size=10)
X, y = generator.next()
```

Initially loading the data still takes time, although it should speed up proportionally to the number of processors you have in your machine. After the generator has been created, calling next() returns batches instantly. Also, this way, you don't run into trouble with exceeding memory.

7.3   *Training a deep-learning model on human game-play data*

Now that you have access to high-dan Go data and processed it to fit a move-prediction model, let's connect the dots and build a deep neural network for this data. In our GitHub repository, you'll find a module called *networks* within our dlgo package that you'll use to provide example architectures of neural networks that you can use as baselines to build strong move-prediction models. For instance, you'll find three convolutional neural networks of varying complexity in the networks module, called small.py, medium.py, and large.py. Each of these files contains a layers function that returns a list of layers that you can add to a sequential Keras model.

You'll build a convolutional neural network consisting of four convolutional layers, followed by a final dense layer, all with ReLU activations. On top of that, you'll use a new utility layer right before each convolutional layer—a ZeroPadding2D layer. Zero padding is an operation in which the input features are *padded* with 0s. Let's say you use your one-plane encoder from chapter 6 to encode the board as a 19×19 matrix. If you specify a padding of 2, that means you add two columns of 0s to the left and right, as well as two rows of 0s to the top and bottom of that matrix, resulting in an enlarged 23×23 matrix. You use zero padding in this situation to artificially increase the input of a convolutional layer, so that the convolution operation doesn't shrink the image by too much.

Before we show you the code, we have to discuss a small technicality. Recall that both input and output of convolutional layers are four-dimensional: we provide a mini-batch of a number of filters that are two-dimensional each (namely, they have width and height). The *order* in which these four dimensions (mini-batch size, number of filters, width, and height) are represented is a matter of convention, and you mainly find two such orderings in practice. Note that filters are also often referred to as channels (C), and the mini-batch size is also called number (N) of examples. Moreover, you can use shorthand for width (W) and height (H). With this notation, the two predominant orderings are NWHC and NCWH. In Keras, this ordering is called data_format, and NWHC is called channels_last, and NCWH channels_first, for somewhat obvious reasons. Now, the way you built your first Go board encoder, the one-plane encoder, is in *channels first* convention (an encoded board has shape 1,19,19, meaning the single encoded plane comes *first*). That means you have to provide data_format=channels_first as an argument to all convolutional layers. Let's have a look at what the model from small.py looks like.

Listing 7.18　Specifying layers for a small convolutional network for Go move prediction

```
from keras.layers.core import Dense, Activation, Flatten
from keras.layers.convolutional import Conv2D, ZeroPadding2D


def layers(input_shape):
    return [
        ZeroPadding2D(padding=3, input_shape=input_shape,
                data_format='channels_first'),
        Conv2D(48, (7, 7), data_format='channels_first'),
        Activation('relu'),

        ZeroPadding2D(padding=2, data_format='channels_first'),
        Conv2D(32, (5, 5), data_format='channels_first'),
        Activation('relu'),

        ZeroPadding2D(padding=2, data_format='channels_first'),
        Conv2D(32, (5, 5), data_format='channels_first'),
        Activation('relu'),

        ZeroPadding2D(padding=2, data_format='channels_first'),
        Conv2D(32, (5, 5), data_format='channels_first'),
        Activation('relu'),

        Flatten(),
        Dense(512),
        Activation('relu'),
    ]
```

> **Use zero padding layers to enlarge input images.**

> **By using channels_first, you specify that the input plane dimension for your features comes first.**

The `layers` function returns a list of Keras layers that you can add one by one to a `Sequential` model. Using these layers, you can now build an application that carries out the first five steps from the overview in figure 7.1—an application that downloads, extracts, and encodes Go data and uses it to train a neural network. For the training part, you'll use the *data generator* you built. But first, let's import some of the essential components of your growing Go machine-learning library. You need a Go data processor, an encoder, and a neural network architecture to build this application.

Listing 7.19　Core imports for building a neural network for Go data

```
from dlgo.data.parallel_processor import GoDataProcessor
from dlgo.encoders.oneplane import OnePlaneEncoder

from dlgo.networks import small
from keras.models import Sequential
from keras.layers.core import Dense
from keras.callbacks import ModelCheckpoint
```

> **With model checkpoints, you can store progress for time-consuming experiments.**

The last of these imports provides a handy Keras tool called `ModelCheckpoint`. Because you have access to a large amount of data for training, completing a full run of training a model for some epochs can take a few hours or even days. If such an experiment fails for some reason, you better have a backup in place. And that's precisely what model checkpoints do for you: they persist a snapshot of your model after each epoch of training. Even if something fails, you can resume training from the last checkpoint.

Next, let's define training and test data. To do so, you first initialize a OnePlane-Encoder that you use to create a GoDataProcessor. With this processor, you can instantiate a training and a testing data generator that you'll use with a Keras model.

Listing 7.20   Creating training and test generators

```
go_board_rows, go_board_cols = 19, 19                              First you create an
num_classes = go_board_rows * go_board_cols                        encoder of board size.
num_games = 100

encoder = OnePlaneEncoder((go_board_rows, go_board_cols))          Then you initialize a
                                                                   Go data processor
processor = GoDataProcessor(encoder=encoder.name())               with it.

generator = processor.load_go_data('train', num_games, use_generator=True)
test_generator = processor.load_go_data('test', num_games, use_generator=True)
```
**From the processor, you create two data
generators, for training and testing.**

As a next step, you define a neural network with Keras by using the layers function from dlgo.networks.small. You add the layers of this small network one by one to a new sequential network, and then finish off by adding a final Dense layer with softmax activation. You then compile this model with categorical cross-entropy loss and train it with SGD.

Listing 7.21   Defining a Keras model from your small layer architecture

```
input_shape = (encoder.num_planes, go_board_rows, go_board_cols)
network_layers = small.layers(input_shape)
model = Sequential()
for layer in network_layers:
    model.add(layer)
model.add(Dense(num_classes, activation='softmax'))
model.compile(loss='categorical_crossentropy', optimizer='sgd',
    metrics=['accuracy'])
```

To train a Keras model with generators works a little bit differently from training with data sets. Instead of calling fit on your model, you now need to call fit_generator, and you also replace evaluate with evaluate_generator. Moreover, the signatures of these methods are slightly different from what you've seen before. Using fit_generator works by specifying a generator, the number of epochs, and the number of training steps per epoch, which you provide with steps_per_epoch. These three arguments provide the bare minimum to train a model. You also want to validate your training process on test data. For this, you provide `validation_data` with your test data generator and specify the number of validation steps per epoch as `validation_steps`. Lastly, you add a `callback` to your model. Callbacks allow you to track and return additional information during the training process. You use callbacks here to hook in the `ModelCheckpoint` utility to store the Keras model after each epoch. As an example, you train a model for five epochs on a batch size of 128.

Listing 7.22 Fitting and evaluating Keras models with generators

You specify a training data generator for your batch size...

```
epochs = 5
batch_size = 128
model.fit_generator(
    generator=generator.generate(batch_size, num_classes),
    epochs=epochs,
    steps_per_epoch=generator.get_num_samples() / batch_size,
    validation_data=test_generator.generate(
      batch_size, num_classes),
    validation_steps=test_generator.get_num_samples() / batch_size,
    callbacks=[
      ModelCheckpoint('../checkpoints/small_model_epoch_{epoch}.h5')
    ])
model.evaluate_generator(
    generator=test_generator.generate(batch_size, num_classes),
    steps=test_generator.get_num_samples() / batch_size)
```

...and the number of training steps per epoch you execute.

An additional generator is used for validation...

...which also needs a number of steps.

After each epoch, you persist a checkpoint of the model.

For evaluation, you also specify a generator and the number of steps.

Note that if you run this code yourself, you should be aware of the time it may take to complete this experiment. If you run this on a CPU, training an epoch might take a few hours. As it happens, the math used in machine learning has a lot in common with the math used in computer graphics. So in some cases, you can move your neural network computation onto your GPU and get a big speedup. Using a GPU for computation will massively speed up computation, usually by one or two orders of magnitude for convolutional neural networks. TensorFlow has extensive support for moving computation onto certain GPUs, if your machine has suitable drivers available.

NOTE If you want to use a GPU for machine learning, an NVIDIA chip with a Windows or Linux OS is the best-supported combination. Other combinations are possible, but you may spend a lot of time fiddling with drivers.

In case you don't want to try this yourself, or just don't want to do this right now, we've precomputed this model for you. Have a look at our GitHub repository to see the five checkpoint models stored in `checkpoints`, one for each completed epoch. Here's the output of that training run (computed on an old CPU on a laptop, to encourage you to get a fast GPU right away):

```
Epoch 1/5
12288/12288 [==============================] - 14053s 1s/step - loss: 3.5514
➡ - acc: 0.2834 - val_loss: 2.5023 - val_acc: 0.6669
Epoch 2/5
12288/12288 [==============================] - 15808s 1s/step - loss: 0.3028
➡ - acc: 0.9174 - val_loss: 2.2127 - val_acc: 0.8294
Epoch 3/5
12288/12288 [==============================] - 14410s 1s/step - loss: 0.0840
➡ - acc: 0.9791 - val_loss: 2.2512 - val_acc: 0.8413
Epoch 4/5
12288/12288 [==============================] - 14620s 1s/step - loss: 0.1113
➡ - acc: 0.9832 - val_loss: 2.2832 - val_acc: 0.8415
Epoch 5/5
12288/12288 [==============================] - 18688s 2s/step - loss: 0.1647
➡ - acc: 0.9816 - val_loss: 2.2928 - val_acc: 0.8461
```

As you can see, after three epochs, you've reached 98% accuracy on training and 84% on test data. This is a massive improvement over the models you computed in chapter 6! It seems that training a larger network on real data paid off: your network learned to predict moves from 100 games almost perfectly, but generalizes reasonably well. You can be more than happy with the 84% validation accuracy. On the other hand, 100 games' worth of moves is still a tiny data set, and you don't know yet how well you'd do on a much larger corpus of games. After all, your goal is to build a strong Go bot that can compete with strong opponents, not to crush a toy data set.

To build a really strong opponent, you need to work with better Go data encoders next. Your one-plane encoder from chapter 6 is a good first guess, but it doesn't capture the complexity that you're dealing with. In section 7.4 you'll learn about two more-sophisticated encoders that will boost your training performance.

7.4   *Building more-realistic Go data encoders*

Chapters 2 and 3 covered the ko rule in Go quite a bit. Recall that this rule exists to prevent infinite loops in games: you can't play a stone that leads to a situation previously on the board. If we give you a random Go board position and you have to decide whether there's a ko going on, you'd have to guess. There's no way of knowing without having seen the sequence leading up to that position. In particular, your one-plane encoder, which encoded black stones as –1, white ones as 1, and empty positions as 0, can't possibly learn anything about ko. This is just one example, but it goes to show that the OnePlaneEncoder you built in chapter 6 is a little too simplistic to capture everything you need to build a strong Go bot.

In this section, we'll provide you with two more elaborate encoders that led to relatively strong move-prediction performance in the literature. The first one we call SevenPlaneEncoder, which consists of the following seven feature planes. Each plane is a 19 × 19 matrix and describes a different set of features:

- The first plane has a 1 for every *white* stone that has precisely *one* liberty, and 0s otherwise.
- The second and third feature planes have a 1 for white stones with two or at least three liberties, respectively.

- The fourth to sixth planes do the same for black stones; they encode black stones with one, two, or at least three liberties.
- The last feature plane marks points that can't be played because of ko with a 1.

Apart from explicitly encoding the concepts of ko, with this set of features you also model liberties and distinguish between black and white stones. Stones with just one liberty have extra tactical significance, because they're at risk of getting captured on the next turn. (Go players say that a stone with just one liberty is in *atari*.) Because the model can "see" this property directly, it's easier for it to pick up on how that affects game play. By creating planes for concepts such as ko and the number of liberties, you give a hint to the model that these concepts are important, without having to explain how or why they're important.

Let's see how you can implement this by extending the base `Encoder` from the encoders module. Save the following code in sevenplane.py.

Listing 7.23   Initializing a simple seven-plane encoder

```
import numpy as np

from dlgo.encoders.base import Encoder
from dlgo.goboard import Move, Point


class SevenPlaneEncoder(Encoder):
    def __init__(self, board_size):
        self.board_width, self.board_height = board_size
        self.num_planes = 7

    def name(self):
        return 'sevenplane'
```

The interesting part is the encoding of the board position, which is done as follows.

Listing 7.24   Encoding game state with a `SevenPlaneEncoder`

```
    def encode(self, game_state):
        board_tensor = np.zeros(self.shape())
        base_plane = {game_state.next_player: 0,
                      game_state.next_player.other: 3}
        for row in range(self.board_height):
            for col in range(self.board_width):
                p = Point(row=row + 1, col=col + 1)
                go_string = game_state.board.get_go_string(p)
                if go_string is None:
                    if
game_state.does_move_violate_ko(game_state.next_player,
                                                Move.play(p)):
                        board_tensor[6][row][col] = 1
```

Encoding moves prohibited by the ko rule

```
        else:
            liberty_plane = min(3, go_string.num_liberties) - 1
            liberty_plane += base_plane[go_string.color]
            board_tensor[liberty_plane][row][col] = 1        ◄──────┐
    return board_tensor                                             │
```

Encoding black and white stones
with 1, 2, or more liberties

To finish this definition, you also need to implement a few convenience methods, to suffice the `Encoder` interface.

Listing 7.25   Implementing all other `Encoder` methods for your seven-plane encoder

```
    def encode_point(self, point):
        return self.board_width * (point.row - 1) + (point.col - 1)

    def decode_point_index(self, index):
        row = index // self.board_width
        col = index % self.board_width
        return Point(row=row + 1, col=col + 1)

    def num_points(self):
        return self.board_width * self.board_height

    def shape(self):
        return self.num_planes, self.board_height, self.board_width


def create(board_size):
    return SevenPlaneEncoder(board_size)
```

Another encoder that we'll discuss here, and point you to the code in GitHub, is an encoder with 11 feature planes that's similar to `SevenPlaneEncoder`. In this encoder, called `SimpleEncoder`, which you can find under simple.py in the encoders module in GitHub, you use the following feature planes:

- The first four feature planes describe black stones with one, two, three, or four liberties.
- The second four planes describe white stones with one, two, three, or four liberties.
- The ninth plane is set to 1 if it's black's turn, and the tenth if it's white's.
- The last feature plane is again reserved for indicating ko.

This encoder with 11 planes is close to the last one, but is more explicit about whose turn it is and more specific about the number of liberties a stone has. Both are great encoders that will lead to notable improvements in model performance.

Throughout chapters 5 and 6, you learned about many techniques that improve your deep-learning models, but one ingredient remained the same for all experiments: you used stochastic gradient descent as the optimizer. Although SGD provides a great baseline, in the next section we'll teach you about *Adagrad* and *Adadelta*, two optimizers that your training process will greatly benefit from.

7.5 *Training efficiently with adaptive gradients*

To further improve performance of your Go move-prediction models, we'll introduce one last set of tools in this chapter—optimizers other than stochastic gradient descent. Recall from chapter 5 that SGD has a fairly simplistic update rule. If for a parameter W you receive a backpropagation error of ΔW and you have a learning rate of α specified, then updating this parameter with SGD simply means computing $W - \alpha \Delta W$.

In many cases, this update rule can lead to good results, but a few drawbacks exist as well. To address them, you can use many excellent extensions to plain SGD.

7.5.1 *Decay and momentum in SGD*

For instance, a widely used idea is to let the learning rate *decay* over time; with every update step you take, the learning rate becomes smaller. This technique usually works well, because in the beginning your network hasn't learned anything yet, and big update steps might make sense to get closer to a minimum of the loss function. But after the training process has reached a certain level, you should make your updates smaller and make only appropriate refinements to the learning process that don't spoil progress. Usually, you specify learning rate decay by a *decay rate*, a percentage by which you'll decrease the next step.

Another popular technique is that of *momentum*, in which a fraction of the last update step is added to the current one. For instance, if W is a parameter vector that you want to update, ∂W is the current gradient computed for W, and if the last update you used was U, then the next update step will be as follows:

$$W \leftarrow W - \alpha \left(\gamma U + (1 - \gamma) \partial W \right)$$

This fraction γ you keep from the last update is called the *momentum term*. If both gradient terms point in roughly the same direction, your next update step gets reinforced (receives momentum). If the gradients point in opposite directions, they cancel each other out and the gradient gets dampened. The technique is called *momentum* because of the similarity of the physical concept by the same name. Think about your loss function as a surface and the parameters lying on that surface as a ball. Then a parameter update describes movement of the ball. Because you're doing gradient descent, you can even think of this as a ball rolling down the surface, by receiving movements one by one. If the last few (gradient) steps all point in the same general direction, the ball will pick up speed and reach its destination, the minimum of the surface, quicker. The momentum technique exploits this analogy.

If you want to use decay, momentum, or both in SGD with Keras, it's as simple as providing the respective rates to an SGD instance. Let's say you want SGD with a learning rate of 0.1, a 1% decay rate, and 90% momentum; you'd do the following.

Listing 7.26   Initializing SGD in Keras with momentum and learning rate decay

```
from keras.optimizers import SGD
sgd = SGD(lr=0.1, momentum=0.9, decay=0.01)
```

7.5.2   *Optimizing neural networks with Adagrad*

Both learning rate decay and momentum do a good job at refining plain SGD, but a few weaknesses still remain. For instance, if you think about the Go board, professionals will almost exclusively play their first few moves on the third to fifth lines of the board, but, without exception, never on the first or second. In the endgame, the situation is somewhat reversed, in that many of the last moves happen at the border of the board. In all deep-learning models you worked with so far, the last layer was a dense layer of board size (here 19×19). Each neuron of this layer corresponds to a position on the board. If you use SGD, with or without momentum or decay, *the same learning rate is used for each of these neurons.* This can be dangerous. Maybe you did a poor job at shuffling the training data, and the learning rate has decayed so much that endgame moves on the first and second line don't get any significant updates anymore—meaning, no learning. In general, you want to make sure that infrequently observed patterns still get large-enough updates, while frequent patterns receive smaller and smaller updates.

To address the problem caused by setting global learning rates, you can use techniques using *adaptive* gradient methods. We'll show you two of these methods: *Adagrad* and *Adadelta.*

In Adagrad, there's no global learning rate. You *adapt the learning rate per parameter.* Adagrad works pretty well when you have a lot of data and patterns in the data can be found only rarely. Both of these criteria apply to our situation: you have a lot of data, and professional Go game play is so complex that certain move combinations occur infrequently in your data set, although they're considered standard play by professionals.

Let's say you have a weight vector W of length l (it's easier to think of vectors here, but this technique applies more generally to tensors as well) with individual entries W_i. For a given gradient ∂W for these parameters, in plain SGD with a learning rate of α, the update rule for each W_i is as follows:

$$W_i \leftarrow W_i - \alpha \partial W_i$$

In Adagrad, you replace α with a term that adapts dynamically for each index i by looking at how much you've updated W_i in the past. In fact, in Adagrad the individual learning rate will be inversely proportional to the previous updates. To be more precise, in Adagrad, you update parameters as follows:

$$W \leftarrow W - \frac{\alpha}{\sqrt{G + \varepsilon}} \cdot \partial W$$

In this equation, ϵ is a small positive value to ensure you're not dividing by 0, and $G_{i,i}$ is the sum of squared gradients W_i received until this point. We write this as $G_{i,i}$

because you can view this term as part of a square matrix G of length l in which all diagonal entries Gj,j have the form we just described and all off-diagonal terms are 0. A matrix of this form is called a *diagonal matrix*. You update G after each parameter update, by adding the latest gradient contributions to the diagonal elements. That's all there is to defining Adagrad, but if you want to write this update rule in a concise form independent of the index i, this is the way to do it:

$$W \leftarrow W - \frac{\alpha}{\sqrt{G + \varepsilon} \cdot \partial W}$$

Note that because G is a matrix, you need to add ϵ to each entry Gi,j and divide α by each such entry. Moreover, by $G \cdot \partial W$ you mean matrix multiplication of G with ∂W. To use Adagrad with Keras, compiling a model with this optimizer works as follows.

Listing 7.27 Using the Adagrad optimizer for Keras models

```
from keras.optimizers import Adagrad
adagrad = Adagrad()
```

A key benefit of Adagrad over other SGD techniques is that you don't have to manually set the learning rate—one thing less to worry about. It's hard enough already to find a good network architecture and tune all the parameters for the model. In fact, you could alter the initial learning rate in Keras by using `Adagrad(lr=0.02)`, but it's not recommended to do so.

7.5.3 *Refining adaptive gradients with Adadelta*

An optimizer that's similar to Adagrad and is an extension of it is *Adadelta*. In this optimizer, instead of accumulating all past (squares of) gradients in G, you use the same idea we've shown you in the momentum technique and keep only a *fraction of the last update* and add the current gradient to it:

$$G \leftarrow \gamma G + (1 - \gamma) \partial W$$

Although this idea is roughly what happens in Adadelta, the details that make this optimizer work and that lead to its precise update rule are a little too intricate to present here. We recommend that you look into the original paper for more details (https://arxiv.org/abs/1212.5701).

In Keras, you use the Adadelta optimizer as follows.

Listing 7.28 Using the Adadelta optimizer for Keras models

```
from keras.optimizers import Adadelta
adadelta = Adadelta()
```

Both Adagrad and Adadelta are hugely beneficial to training deep neural networks on Go data, as compared to stochastic gradient descent. In later chapters, you'll often use one or the other as an optimizer in more-advanced models.

7.6   *Running your own experiments and evaluating performance*

Throughout chapters 5, 6, and this one, we've shown you many deep-learning techniques. We gave you some hints and sample architectures that made sense as a baseline, but now it's time to train your own models. In machine-learning experiments, it's crucial to try various combinations of *hyperparameters*, such as the number of layers, which layers to choose, how many epochs to train for, and so on. In particular, with deep neural networks, the number of choices you face can be overwhelming. It's not always as clear how tweaking a specific knob impacts model performance. Deep-learning researchers can rely on a large corpus of experimental results and further theoretical arguments from decades of research to back their intuition. We can't provide you with that deep a level of knowledge here, but we can help get you started building intuition of your own.

A crucial factor in achieving strong results in experimental setups such as ours—namely, training a neural network to predict Go moves as well as possible—is a *fast experimentation cycle*. The time it takes you to build a model architecture, start model training, observe and evaluate performance metrics, and then go back to adjust your model and start the process anew has to be short. When you look at data science challenges such as those hosted on kaggle.com, it's often the teams *who tried the most* that win. Luckily for you, Keras was built with fast experimentation in mind. It's also one of the prime reasons we chose it as deep-learning framework for this book. We hope you agree that you can build neural networks with Keras quickly and that changing your experimental setup comes naturally.

7.6.1   *A guideline to testing architectures and hyperparameters*

Let's have a look at a few practical considerations when building a move-prediction network:

- Convolutional neural networks are a good candidate for Go move-prediction networks. Make sure to convince yourself that working with only dense layers will result in inferior prediction quality. Building a network that consists of several convolutional layers and one or two dense layers at the end is usually a must. In later chapters, you'll see more-complex architectures, but for now, work with convolutional networks.

- In your convolutional layers, vary the kernel sizes to see how this change influences model performance. As a rule of thumb, kernel sizes between 2 and 7 are suitable, and you shouldn't go much larger than that.

- If you use pooling layers, make sure to experiment with both max and average pooling, but more important, don't choose a too large pooling size. A practical upper bound might be 3 in your situation. You may also want to try building networks without pooling layers, which might be computationally more expensive, but can work pretty well.

- Use dropout layers for regularization. In chapter 6, you saw how dropout can be used to prevent your model from overfitting. Your networks will generally benefit from adding in dropout layers, as long as you don't use too many of them and don't set the dropout rate too high.

- Use softmax activation in your last layer for its benefit of producing probability distributions and use it in combination with categorical cross-entropy loss, which suits your situation very well.

- Experiment with different activation functions. We've introduced you to ReLU, which should act as your default choice for now, and sigmoid activations. You can use plenty of other activation functions in Keras, such as elu, selu, PReLU, and LeakyReLU. We can't discuss these ReLU variants here, but their usage is well described at https://keras.io/activations/.

- Varying mini-batch size has an impact on model performance. In prediction problems such as MNIST from chapter 5, it's usually recommended to choose mini-batches in the same order of magnitude as the number of classes. For MNIST, you often see mini-batch sizes ranging from 10 to 50. If data is perfectly randomized, this way, each gradient will receive information from each class, which makes SGD generally perform better. In our use case, some Go moves are played much more often than others. For instance, the four corners of the board are rarely played, especially compared with the star points. We call this a *class imbalance* in our data. In this case, you can't expect a mini-batch to cover all the classes, and should work with mini-batch sizes ranging from 16 to 256 (which is what you find in the literature).

The choice of optimizer also has a considerable impact on how well your network learns. SGD with or without learning rate decay, as well as Adagrad and Adadelta, already give you options to experiment with. Under https://keras.io/optimizers/ you'll find other optimizers that your model training process might benefit from.

The number of epochs used to train a model has to be chosen appropriately. If you use model checkpointing and track various performance metrics per epoch, you can effectively measure when training stops improving. In the next and final section of this chapter, we briefly discuss how to evaluate performance metrics. As a general rule of thumb, given enough compute power, set the number of epochs too high rather than too low. If model training stops improving or even gets worse through overfitting, you can still take an earlier checkpoint model for your bot.

Weight initializers

Another crucial aspect for tuning deep neural networks is how to initialize the weights before training starts. Because optimizing a network means finding a set of weights corresponding to a minimum on the loss surface, the weights you start with are important. In your network implementation from chapter 5, you *randomly* assigned initial weights, which is generally a bad idea.

Weight initializations are an interesting topic of research and almost deserve a chapter of their own. Keras has many weight initialization schemes, and each layer with weights can be initialized accordingly. The reason you don't cover them in the main text is that the initializers Keras chooses by default are usually so good that it's not worth bothering to change them. Usually, it's other aspects of your network definition that require attention. But it's good to know that there are differences, and advanced users might want to experiment with Keras initializers, found at https://keras.io/initializers/, as well.

7.6.2   *Evaluating performance metrics for training and test data*

In section 7.3, we showed you results of a training run performed on a small data set. The network we used was a relatively small convolutional network, and we trained this network for five epochs. In this experiment, we tracked loss and accuracy on training data and used test data for validation. At the end, we computed accuracy on test data. That's the general workflow you should follow, but how do you judge when to stop training or detect when something is off? Here are a few guidelines:

- Your training accuracy and loss should generally improve for each epoch. In later epochs, these metrics will taper off and sometimes fluctuate a little. If you don't see any improvement for a few epochs, you might want to stop.
- At the same time, you should see what your validation loss and accuracy look like. In early epochs, validation loss will drop consistently, but in later epochs, what you often see is that it plateaus and often starts to increase again. This is a sure sign that the network starts to overfit on the training data.
- If you use model checkpointing, pick the model from the epoch with high training accuracy that still has a low validation error.
- If both training and validation loss are high, try to choose a deeper network architecture or other hyperparameters.
- In case your training error is low, but validation error high, your model is overfitting. This scenario usually doesn't occur when you have a truly large training data set. With more than 170,000 Go games and many million moves to learn from, you should be fine.
- Choose a training data size that makes sense for your hardware requirements. If training an epoch takes more than a few hours, it's just not that much fun. Instead, try to find a well-performing model among many tries on a medium-sized data set and then train this model once again on the largest data set possible.
- If you don't have a good GPU, you might want to opt for training your model in the cloud. In appendix D, we'll show you how to train a model on a GPU using Amazon Web Services (AWS).
- When comparing runs, don't stop a run that looks worse than a previous run too early. Some learning processes are slower than others—and might eventually catch up or even outperform other models.

You might ask yourself how strong a bot you can potentially build with the methods presented in this chapter. A theoretical upper bound is this: the network can never get better at playing Go than the data you feed it. In particular, using just supervised deep-learning techniques, as you did in the last three chapters, won't surpass human game play. In practice, with enough compute power and time, it's definitely possible to reach results up to about 2 dan level.

To reach super-human performance of game play, you need to work with *reinforcement-learning* techniques, introduced in chapters 9 to 12. Afterward, you can combine tree search from chapter 4, reinforcement learning, and supervised deep learning to build even stronger bots in chapters 13 and 14.

But before you go deeper into the methodology of building stronger bots, in the next chapter we'll show you how to *deploy* a bot and let it interact with its environment by playing against either human opponents or other bots.

7.7 Summary

- The ubiquitous Smart Game Format (SGF) for Go and other game records is useful to build data for neural networks.
- Go data can be processed in parallel for speed and efficiently represented as generators.
- With strong amateur-to-professional game records, you can build deep-learning models that predict Go moves quite well.
- If you know certain properties of your training data that are important, you can explicitly encode them in *feature planes*. Then the model can quickly learn connections between the feature planes and the results you're trying to predict. For a Go bot, you can add feature planes that represent concepts such as the number of liberties (adjacent empty points) a string of stones has.
- You can train more efficiently by using adaptive gradient techniques such as Adagrad or Adadelta. These algorithms adjust the learning rate on the fly as training progresses.
- End-to-end model training can be achieved in a relatively small script that you can use as a template for your own experiments.

Deploying
bots in the wild

This chapter covers

- Building an end-to-end application to train and run a Go bot
- Running a frontend to play against your bot
- Letting your bot play against other bots locally
- Deploying your bot on an online Go server

By now, you know how to build and train a strong deep-learning model for Go move prediction—but how do you integrate this into an application that plays games against opponents? Training a neural network is just one part of building an end-to-end application, whether you're playing yourself or letting your bot compete against other bots. The trained model has to be integrated into an engine that can be played against.

In this chapter, you'll build a simple Go model server and two frontends. First, we provide you with an HTTP frontend that you can use to play against your bot. Then, we introduce you to the Go Text Protocol (GTP), a widely used protocol that Go bots use to exchange information, so your bot can play against other bots like

GNU Go or *Pachi*, two freely available Go programs based on GTP. Finally, we show you how to deploy your Go bot on Amazon Web Services (AWS) and connect it against the Online Go Server (OGS). Doing so will allow your bots to play ranked games in a real environment, compete against other bots and human players worldwide, and even enter tournaments. To do all this, we'll show you how to tackle the following tasks:

- *Building a move-prediction agent*—The neural networks you trained in chapters 6 and 7 need to be integrated into a framework that allows you to use them in game play. In section 8.1, we'll pick up the idea of *agents* from chapter 3 (in which you created a randomly playing agent) as a basis to serve a deep-learning bot.
- *Providing a graphical interface*—As humans, to conveniently play against a Go bot, we need some sort of (graphical) interface. Although so far we've been happy with command-line interfaces, in section 8.2 we'll equip you with a fun-to-play frontend for your bot.
- *Deploying a bot in the cloud*—If you don't have a powerful GPU in your computer, you won't get far training strong Go bots. Luckily, most big cloud providers offer GPU instances on demand. But even if you have a strong-enough GPU for training, you still may want to host your previously trained model on a server. In section 8.3, we'll show you how this can be done and refer you to appendix D for more details on how to set everything up in AWS.
- *Talking to other bots*—Humans use graphical and other interfaces to interact with each other. For bots, it's customary to communicate through a standardized protocol. In section 8.4, we'll introduce you to the common Go Text Protocol (GTP). This is an essential component for the following two points:
  - *Playing against other bots*—You'll then build a GTP frontend for your bot to let it play against other programs in section 8.5. We'll show you how to let your bot play against two other Go programs locally, to see how well your creation does.
  - *Deploying a bot on an online Go server*—In section 8.6, we'll finally show you how to deploy a bot on an online Go platform so that registered users and other bots can compete against your bot. This way, your bot can even enter ranked games and enter tournaments, all of which we'll show you in this last section. Because most of this material is technical, you'll find most of the details in appendix E.

8.1 Creating a move-prediction agent from a deep neural network

Now that you have all the building blocks in place to build a strong neural network for Go data, let's integrate such networks into an *agent* that will serve them. Recall from chapter 3 the concept of `Agent`. We defined it as a class that can select the next move for the current game state, by implementing a `select_move` method. Let's write a `DeepLearningAgent` by using Keras models and our Go board `Encoder` concept (put this code into predict.py in the agent module in dlgo).

Listing 8.1 Initializing an agent with a Keras model and a Go board encoder

```python
import numpy as np

from dlgo.agent.base import Agent
from dlgo.agent.helpers import is_point_an_eye
from dlgo import encoders
from dlgo import goboard
from dlgo import kerasutil


class DeepLearningAgent(Agent):
    def __init__(self, model, encoder):
        Agent.__init__(self)
        self.model = model
        self.encoder = encoder
```

You'll use the encoder to transform the board state into features, and you'll use the model to predict the next move. In fact, you'll use the model to compute a whole probability distribution of possible moves that you'll later sample from.

Listing 8.2 Encoding board state and predicting move probabilities with a model

```python
    def predict(self, game_state):
        encoded_state = self.encoder.encode(game_state)
        input_tensor = np.array([encoded_state])
        return self.model.predict(input_tensor)[0]

    def select_move(self, game_state):
        num_moves = self.encoder.board_width * self.encoder.board_height
        move_probs = self.predict(game_state)
```

Next, you alter the probability distribution stored in move_probs a little. First, you compute the cube of all values to drastically increase the distance between more-likely and less-likely moves. You want the best possible moves to be picked much more often. Then you use a trick called *clipping* that prevents move probabilities from being too close to either 0 or 1. This is done by defining a small positive value, $\varepsilon = 0.000001$, and setting values smaller than ε to ε, and values larger than $1 - \varepsilon$ to $1 - \varepsilon$. Afterward, you normalize the resulting values to end up with a probability distribution once again.

Listing 8.3 Scaling, clipping, and renormalizing your move probability distribution

Prevents move probabilities from getting stuck at 0 or 1

Increases the distance between the more likely and least likely moves

```python
        move_probs = move_probs ** 3
        eps = 1e-6
        move_probs = np.clip(move_probs, eps, 1 - eps)
        move_probs = move_probs / np.sum(move_probs)
```

Renormalizes to get another probability distribution

You do this transformation because you want to sample moves from this distribution, according to their probabilities. Instead of sampling moves, another viable strategy would be to always take the most likely move (taking the maximum over the distribution). The benefit of the way you're doing it is that sometimes other moves get chosen, which might be especially useful when there isn't one single move that sticks out from the rest.

Listing 8.4   Trying to apply moves from a ranked candidate list

```
                    candidates = np.arange(num_moves)                    Turns the
Samples             ranked_moves = np.random.choice(                     probabilities
potential               candidates, num_moves, replace=False, p=move_probs)   into a ranked
candidates          for point_idx in ranked_moves:                       list of moves
                        point = self.encoder.decode_point_index(point_idx)
                        if game_state.is_valid_move(goboard.Move.play(point)) and \
                            not is_point_an_eye(game_state.board, point,
                    game_state.next_player):                              Starting from the top,
                            return goboard.Move.play(point)               finds a valid move that
                    return goboard.Move.pass_turn()                       doesn't reduce eye-space

                    If no legal and non-self-destructive
                    moves are left, passes
```

For convenience, you also want to persist a `DeepLearningAgent`, so you can pick it up at a later point. The prototypical situation in practice is this: you train a deep-learning model and create an agent, which you then persist. At a later point, this agent gets deserialized and served, so human players or other bots can play against it. To do the serialization step, you hijack the serialization format of Keras. When you persist a Keras model, it gets stored in HDF5, an efficient serialization format. HDF5 files contain flexible *groups* that are used to store *meta-information* and *data*. For any Keras model, you can call `model.save("model_path.h5")` to persist the full model, meaning the neural network architecture and all weights, to the local file model_path.h5. The only thing you need to do before persisting a Keras model like this is to install the Python library h5py; for instance, with `pip install h5py`.

To store a complete agent, you can add an additional group for information about your Go board encoder.

Listing 8.5   Serializing a deep-learning agent

```
def serialize(self, h5file):
    h5file.create_group('encoder')
    h5file['encoder'].attrs['name'] = self.encoder.name()
    h5file['encoder'].attrs['board_width'] = self.encoder.board_width
    h5file['encoder'].attrs['board_height'] = self.encoder.board_height
    h5file.create_group('model')
    kerasutil.save_model_to_hdf5_group(self.model, h5file['model'])
```

Finally, after you serialize a model, you also need to know how to load it from an HDF5 file.

Listing 8.6 Deserializing a `DeepLearningAgent` from an HDF5 file

```
def load_prediction_agent(h5file):
    model = kerasutil.load_model_from_hdf5_group(h5file['model'])
    encoder_name = h5file['encoder'].attrs['name']
    if not isinstance(encoder_name, str):
        encoder_name = encoder_name.decode('ascii')
    board_width = h5file['encoder'].attrs['board_width']
    board_height = h5file['encoder'].attrs['board_height']
    encoder = encoders.get_encoder_by_name(
        encoder_name, (board_width, board_height))
    return DeepLearningAgent(model, encoder)
```

This completes our definition of a deep-learning agent. As a next step, you have to make sure this agent connects and interacts with an environment. You do this by embedding `DeepLearningAgent` into a web application that human players can play against in their browser.

8.2 *Serving your Go bot to a web frontend*

In chapters 6 and 7, you designed and trained a neural network that predicts what move a human would play in a Go game. In section 8.1, you turned that model for move *prediction* into a `DeepLearningAgent` that does move *selection*. The next step is to play your bot! Back in chapter 3, you built a bare-bones interface in which you could type in moves on your keyboard, and your benighted `RandomBot` would print its reply to the console. Now that you've built a more sophisticated bot, it deserves a nicer frontend to communicate moves with a human player.

In this section, you'll connect the `DeepLearningAgent` to a Python web application, so you can play against it in your web browser. You'll use the lightweight Flask library to serve such an agent via HTTP. On the browser side, you'll use a JavaScript library called jgoboard to render a Go board that humans can use. The code can be found in our repository on GitHub, in the httpfrontend module in dlgo. We don't explicitly discuss this code here, because we don't want to distract from the main topic, building a Go AI, by digressing into web development techniques in other languages (such as HTML or JavaScript). Instead, we'll give you an overview of what the application does and how to use it in an end-to-end example. Figure 8.1 provides an overview of the application you're going to build in this chapter.

If you look into the structure of httpfrontend, you find a file called server.py that has a single, well-documented method, get_web_app, that you can use to return a web

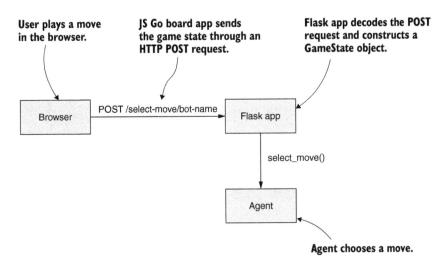

Figure 8.1 Building a web frontend for your Go bot. The httpfrontend module starts a Flask web server that decodes HTTP requests and passes them to one or more Go-playing agents. In the browser, a client based on the jgoboard library communicates with the server over HTTP.

application to run. Here's an example of how to use get_web_app to load a random bot and serve it.

Listing 8.7 Registering a random agent and starting a web application with it

```
from dlgo.agent.naive import RandomBot
from dlgo.httpfrontend.server import get_web_app

random_agent = RandomBot()
web_app = get_web_app({'random': random_agent})
web_app.run()
```

When you run this example, a web application will start on localhost (127.0.0.1), listening on port 5000, which is the default port used in Flask applications. The Random-Bot you just registered as random corresponds to an HTML file in the static folder in httpfrontend: play_random_99.html. In this file, a Go board is rendered, and it's also the place in which the rules of human-bot game play are defined. The human opponent starts with the black stones; the bot takes white. Whenever a human move has been played, the route/select-move/random is triggered to receive the next move from the bot. After the bot move has been received, it's applied to the board, and it's the human's move once again. To play against this bot, navigate to http://127.0.0.1:5000/static/play_random_99.html in your browser. You should see a playable demo, as shown in figure 8.2.

You'll add more and more bots in the next chapters, but for now note that another frontend is available under play _predict_19.html. This web frontend talks to a bot called predict and can be used to play 19 × 19 games. Therefore, if you train a Keras neural network model on Go data and use a Go board encoder, you can first create an instance agent = DeepLearningAgent(model, encoder) and then register it in a web application web_app = get_web_app({'predict': agent}) that you can then start with web_app.run().

8.2.1 *An end-to-end Go bot example*

Figure 8.3 shows an end-to-end example covering the whole process (the same flow we introduced in the begin-

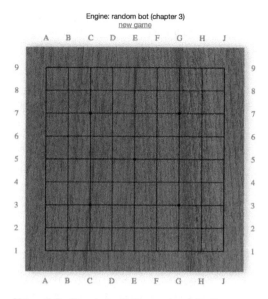

Figure 8.2   Running a Python web application to play against a Go bot in your browser

ning of chapter 7). You start with the imports you need and load Go data into features *X* and labels *y* by using an encoder and a Go data processor, as shown in listing 8.8.

Listing 8.8   Loading features and labels from Go data with a processor

```
import h5py

from keras.models import Sequential
from keras.layers import Dense

from dlgo.agent.predict import DeepLearningAgent, load_prediction_agent
from dlgo.data.parallel_processor import GoDataProcessor
from dlgo.encoders.sevenplane import SevenPlaneEncoder
from dlgo.httpfrontend import get_web_app
from dlgo.networks import large

go_board_rows, go_board_cols = 19, 19
nb_classes = go_board_rows * go_board_cols
encoder = SevenPlaneEncoder((go_board_rows, go_board_cols))
processor = GoDataProcessor(encoder=encoder.name())

X, y = processor.load_go_data(num_samples=100)
```

Equipped with features and labels, you can build a deep convolutional neural network and train it on this data. This time, you choose the large network from dlgo.networks and use Adadelta as the optimizer.

Listing 8.9   Building and running a large Go move-predicting model with Adadelta

```
input_shape = (encoder.num_planes, go_board_rows, go_board_cols)
model = Sequential()
```

```
network_layers = large.layers(input_shape)
for layer in network_layers:
    model.add(layer)
model.add(Dense(nb_classes, activation='softmax'))
model.compile(loss='categorical_crossentropy', optimizer='adadelta',
    metrics=['accuracy'])

model.fit(X, y, batch_size=128, epochs=20, verbose=1)
```

After the model has finished training, you can create a Go bot from it and save this bot in HDF5 format.

Listing 8.10 Creating and persisting a `DeepLearningAgent`

```
deep_learning_bot = DeepLearningAgent(model, encoder)
deep_learning_bot.serialize("../agents/deep_bot.h5")
```

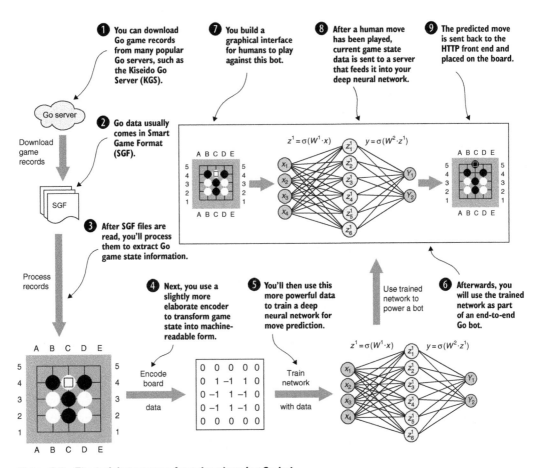

Figure 8.3 The training process for a deep-learning Go bot

Finally, you can load the bot from file and serve it in a web application.

> **Listing 8.11   Loading a bot back into memory and serving it in a web application**

```
model_file = h5py.File("../agents/deep_bot.h5", "r")
bot_from_file = load_prediction_agent(model_file)

web_app = get_web_app({'predict': bot_from_file})
web_app.run()
```

Of course, if you've already trained a strong bot, you can skip all but the last part. For instance, you could load one of the models stored in checkpoints in chapter 7 and see how they perform as opponents in action by changing the model_file accordingly.

8.3   *Training and deploying a Go bot in the cloud*

Until this point, all development took place on your local machine at home. If you're in the good position to have a modern GPU available on your computer, training the deep neural networks we developed in chapters 5–7 isn't of concern for you. If you don't have a powerful GPU or can't spare any compute time on it, it's usually a good option to *rent compute time on a GPU in the cloud.*

If you disregard training for now and assume you have a strong bot already, serving this bot is another situation in which cloud providers can come in handy. In section 8.2, you ran a bot via a web application hosted from localhost. If you want to share your bot with friends or make it public, that's not exactly ideal. You neither want to ensure that your computer runs night and day, nor give the public access to your machine. By hosting your bot in the cloud, you separate development from deployment and can simply share a URL with anyone who's interested in playing your bot.

Because this topic is important, but somewhat special and only indirectly related to machine learning, we entirely outsourced it to appendix D. Reading and applying the techniques from this appendix is entirely optional, but recommended. In appendix D, you'll learn how to get started with one particular cloud provider, Amazon Web Services (AWS). You'll learn the following skills in the appendix:

- Creating an account with AWS
- Flexibly setting up, running, and terminating virtual server instances
- Creating an AWS instance suitable for deep-learning model training on a cloud GPU at reasonable cost
- Deploying your Go bot served over HTTP on an (almost) free server

On top of learning these useful skills, appendix D is also a prerequisite for deploying a full-blown Go bot that connects to an online Go server, a topic we cover later in section 8.6.

8.4   *Talking to other bots: the Go Text Protocol*

In section 8.2, you saw how to integrate your bot framework into a web frontend. For this to work, you handled communication between the bot and human player with the

Hypertext Transfer Protocol (HTTP), one of the core protocols running the web. To avoid distraction, we purposefully left out all the details, but having a *standardized protocol* in place is necessary to pull this off. Humans and bots don't share a common language to exchange Go moves, but a protocol can act as a bridge.

The Go Text Protocol (GTP) is the de facto standard used by Go servers around the world to connect humans and bots on their platforms. Many offline Go programs are based on GTP as well. This section introduces you to GTP by example; you'll implement part of the protocol in Python and use this implementation to let your bots play against other Go programs.

In appendix C, we explain how to install GNU Go and Pachi, two common Go programs available for practically all operating systems. We recommend installing both, so please make sure to have both programs on your system. You don't need any frontends, just the plain command-line tools. If you have GNU Go installed, you can start it in GTP mode by running the following:

```
gnugo --mode gtp
```

Using this mode, you can now explore how GTP works. GTP is a text-based protocol, so you can type commands into your terminal and hit Enter. For instance, to set up a 9 × 9 board, you can type boardsize 9. This will trigger GNU Go to return a response and acknowledge that the command has been executed correctly. Every successful GTP command triggers a response starting with the symbol =, whereas failed commands lead to a ?. To check the current board state, you can issue the command showboard, which will print out an empty 9 × 9 board, as expected.

In actual game play, two commands are the most important: genmove and play. The first command, genmove, is used to ask a GTP bot to generate the next move. The GTP bot will usually also apply this move to its game state internally. All this command needs as arguments is the player color, either black or white. For instance, to generate a white move and place it on GNU Go's board, type genmove white. This will lead to a response such as = C4, meaning GNU Go accepts this command (=) and places a white stone at C4. As you can see, GTP accepts standard coordinates as introduced in chapters 2 and 3.

The other game-play relevant move for us is play. This command is used to let a GTP bot know it has to play a move on the board. For instance, you could tell GNU Go that you want it to play a black move on D4 by issuing play black D4, which will return an = to acknowledge this command. When two bots play against each other, they'll take turns asking each other to genmove the next move, and then play the move from the response on their own board. This is all pretty straightforward—but we left out many details. A complete GTP client has a lot more commands to handle, ranging from handling handicap stones to managing time settings and counting rules. If you're interested in the details of GTP, see http://mng.bz/MWNQ. Having said that, at a basic level genmove and play will be enough to let your deep-learning bots play against GNU Go and Pachi.

To handle GTP and wrap your `Agent` concept so it can exchange Go moves by using this protocol, you create a new dlgo module called gtp. You can still try to follow the implementation alongside this main text, but from this chapter on, we suggest directly following our implementation on GitHub at http://mng.bz/a4Wj.

To start, let's formalize what a GTP command is. To do so, we have to note that on many Go servers, commands get a sequence number to make sure that we can match commands and responses. These sequence numbers are optional and can be `None`. For us, a GTP command consists of a sequence number, a command, and potentially multiple arguments to that command. You place this definition in command.py in the gtp module.

> **Listing 8.12   Python implementation of a GTP command**

```python
class Command:
    def __init__(self, sequence, name, args):
        self.sequence = sequence
        self.name = name
        self.args = tuple(args)

    def __eq__(self, other):
        return self.sequence == other.sequence and \
            self.name == other.name and \
            self.args == other.args

    def __repr__(self):
        return 'Command(%r, %r, %r)' % (self.sequence, self.name, self.args)

    def __str__(self):
        return repr(self)
```

Next, you want to parse text input from the command line into `Command`. For instance, parsing "999 play white D4" should result in `Command(999, 'play', ('white', 'D4'))`. The parse function used for this goes into command.py as well.

> **Listing 8.13   Parsing a GTP `Command` from plain text**

```python
def parse(command_string):
    pieces = command_string.split()
    try:
        sequence = int(pieces[0])      ⟵┐ GTP commands may start with
        pieces = pieces[1:]             │ an optional sequence number.
    except ValueError:          ⟵┐
        sequence = None          │ If the first piece isn't numeric,
    name, args = pieces[0], pieces[1:]  there's no sequence number.
    return Command(sequence, name, args)
```

We've just argued that GTP coordinates come in standard notation, so parsing GTP coordinates into `Board` positions and vice versa is simple. You define two helper functions to convert between coordinates and positions in board.py within gtp.

Listing 8.14 Converting between GTP coordinates and your internal `Point` type

```
from dlgo.gotypes import Point
from dlgo.goboard_fast import Move

def coords_to_gtp_position(move):
    point = move.point
    return COLS[point.col - 1] + str(point.row)

def gtp_position_to_coords(gtp_position):
    col_str, row_str = gtp_position[0], gtp_position[1:]
    point = Point(int(row_str), COLS.find(col_str.upper()) + 1)
    return Move(point)
```

8.5   *Competing against other bots locally*

Now that you understand the basics of GTP, let's dive right into an application and build a program that loads one of your bots and lets it compete against either GNU Go or Pachi. Before we present this program, we have just one technicality left to resolve—when our bot should resign a game or pass.

8.5.1   *When a bot should pass or resign*

At the current development status, your deep-learning bots have no means of knowing when to stop playing. The way you designed them so far, your bot will always pick the best move to play. This can be detrimental toward the end of the game, when it might be better to pass or even resign when the situation looks a little too bad. For this reason, you'll impose *termination strategies*: you'll explicitly tell the bot when to stop. In chapters 13 and 14, you'll learn powerful techniques that'll render this entirely useless (your bot will learn to judge the current board situation and thereby learn that sometimes it's best to stop). But for now, this concept is useful and will help you on the way to deploy a bot against other opponents.

You build the following `TerminationStrategy` in a file called termination.py in the agent module of dlgo. All it does is decide when you should pass or resign—and by default, you never pass or resign.

Listing 8.15 A termination strategy tells your bot when to end a game

```
from dlgo import goboard
from dlgo.agent.base import Agent
from dlgo import scoring

class TerminationStrategy:

    def __init__(self):
        pass

    def should_pass(self, game_state):
        return False

    def should_resign(self, game_state):
        return False
```

A simple heuristic for stopping game play is to pass when your opponent passes. You have to rely on the fact that your opponent knows when to pass, but it's a start, and it works well against GNU Go and Pachi.

Listing 8.16   Passing whenever an opponent passes

```
class PassWhenOpponentPasses(TerminationStrategy):

    def should_pass(self, game_state):
        if game_state.last_move is not None:
            return True if game_state.last_move.is_pass else False

def get(termination):
    if termination == 'opponent_passes':
        return PassWhenOpponentPasses()
    else:
        raise ValueError("Unsupported termination strategy: {}"
                         .format(termination))
```

In termination.py, you also find another strategy called `ResignLargeMargin` that resigns whenever the estimated score of the game goes too much in favor of the opponent. You can cook up many other such strategies, but keep in mind that ultimately you can get rid of this crutch with machine learning.

The last thing you need in order to let bots play against each other is to equip an `Agent` with a `TerminationStrategy` so as to pass and resign when appropriate. This `TerminationAgent` class goes into termination.py as well.

Listing 8.17   Wrapping an agent with a termination strategy

```
class TerminationAgent(Agent):

    def __init__(self, agent, strategy=None):
        Agent.__init__(self)
        self.agent = agent
        self.strategy = strategy if strategy is not None \
            else TerminationStrategy()

    def select_move(self, game_state):
        if self.strategy.should_pass(game_state):
            return goboard.Move.pass_turn()
        elif self.strategy.should_resign(game_state):
            return goboard.Move.resign()
        else:
            return self.agent.select_move(game_state)
```

8.5.2   *Let your bot play against other Go programs*

Having discussed termination strategies, you can now turn to pairing your Go bots with other programs. Under play_local.py in the gtp module, find a script that sets up a game between one of your bots and either GNU Go or Pachi. Go through this script step-by-step, starting with the necessary imports.

Listing 8.18 Imports for your local bot runner

```
import subprocess
import re
import h5py

from dlgo.agent.predict import load_prediction_agent
from dlgo.agent.termination import PassWhenOpponentPasses, TerminationAgent
from dlgo.goboard_fast import GameState, Move
from dlgo.gotypes import Player
from dlgo.gtp.board import gtp_position_to_coords, coords_to_gtp_position
from dlgo.gtp.utils import SGFWriter
from dlgo.utils import print_board
from dlgo.scoring import compute_game_result
```

You should recognize most of the imports, with the exception of SGFWriter. This is a little utility class from dlgo.gtp.utils that keeps track of the game and writes an SGF file at the end.

To initialize your game runner LocalGtpBot, you need to provide a deep-learning agent and optionally a termination strategy. Also, you can specify how many handicap stones should be used and which bot opponent should be played against. For the latter, you can choose between gnugo and pachi. LocalGtpBot will initialize either one of these programs as subprocesses, and both your bot and its opponent will communicate over GTP.

Listing 8.19 Initializing a runner to clash two bot opponents

```
class LocalGtpBot:

    def __init__(self, go_bot, termination=None, handicap=0,          ← You initialize a bot
                 opponent='gnugo', output_sgf="out.sgf",                from an agent and
                 our_color='b'):                                        a termination
        self.bot = TerminationAgent(go_bot, termination)      ←         strategy.
        self.handicap = handicap
        self._stopped = False                          At the end, you write the
        self.game_state = GameState.new_game(19)       game to the provided file
        self.sgf = SGFWriter(output_sgf)          ←    in SGF format.

        self.our_color = Player.black if our_color == 'b' else Player.white
        self.their_color = self.our_color.other

        cmd = self.opponent_cmd(opponent)
        pipe = subprocess.PIPE            You read and write
        self.gtp_stream = subprocess.Popen(   GTP commands from
            cmd, stdin=pipe, stdout=pipe   ←  the command line.
        )

    @staticmethod
    def opponent_cmd(opponent):
        if opponent == 'gnugo':
            return ["gnugo", "--mode", "gtp"]
        elif opponent == 'pachi':
```

You play until the game is stopped by one of the players.

Your opponent will either be GNU Go or Pachi.

```
            return ["pachi"]
        else:
            raise ValueError("Unknown bot name {}".format(opponent))
```

One of the main methods used in the tool we're demonstrating here is `command_and_response`, which sends out a GTP command and reads back the response for this command.

Listing 8.20   Sending a GTP command and receiving a response

```
    def send_command(self, cmd):
        self.gtp_stream.stdin.write(cmd.encode('utf-8'))

    def get_response(self):
        succeeded = False
        result = ''
        while not succeeded:
            line = self.gtp_stream.stdout.readline()
            if line[0] == '=':
                succeeded = True
                line = line.strip()
                result = re.sub('^= ?', '', line)
        return result

    def command_and_response(self, cmd):
        self.send_command(cmd)
        return self.get_response()
```

Playing a game works as follows:

1 Set up the board with the GTP `boardsize` command. You allow only 19 × 19 boards here, because your deep-learning bots are tailored to that.
2 Set the right handicap in the `set_handicap` method.
3 Play the game itself, which you'll cover in the `play` method.
4 Persist the game record as an SGF file.

Listing 8.21   Set up the board, let the opponents play the game, and persist it

```
    def run(self):
        self.command_and_response("boardsize 19\n")
        self.set_handicap()
        self.play()
        self.sgf.write_sgf()

    def set_handicap(self):
        if self.handicap == 0:
            self.command_and_response("komi 7.5\n")
            self.sgf.append("KM[7.5]\n")
        else:
            stones = self.command_and_response("fixed_handicap
    {}\n".format(self.handicap))
            sgf_handicap = "HA[{}]AB".format(self.handicap)
```

```
                    for pos in stones.split(" "):
                        move = gtp_position_to_coords(pos)
                        self.game_state = self.game_state.apply_move(move)
                        sgf_handicap = sgf_handicap + "[" +
        self.sgf.coordinates(move) + "]"
                    self.sgf.append(sgf_handicap + "\n")
```

The game-play logic for your bot clash is simple: while none of the opponents stop, take turns and continue to play moves. The bots do that in methods called play_our _move and play_their_move, respectively. You also clear the screen, and print out the current board situation and a crude estimate of the outcome.

Listing 8.22   A game ends when an opponent signals to stop it

```
def play(self):
    while not self._stopped:
        if self.game_state.next_player == self.our_color:
            self.play_our_move()
        else:
            self.play_their_move()
        print(chr(27) + "[2J")
        print_board(self.game_state.board)
        print("Estimated result: ")
        print(compute_game_result(self.game_state))
```

Playing moves for your bot means asking it to generate a move with select_move, applying it to your board, and then translating the move and sending it over GTP. This needs special treatment for passing and resigning.

Listing 8.23   Asking your bot to generate and play a move that's translated into GTP

```
    def play_our_move(self):
        move = self.bot.select_move(self.game_state)
        self.game_state = self.game_state.apply_move(move)

        our_name = self.our_color.name
        our_letter = our_name[0].upper()
        sgf_move = ""
        if move.is_pass:
            self.command_and_response("play {} pass\n".format(our_name))
        elif move.is_resign:
            self.command_and_response("play {} resign\n".format(our_name))
        else:
            pos = coords_to_gtp_position(move)
            self.command_and_response("play {} {}\n".format(our_name, pos))
            sgf_move = self.sgf.coordinates(move)
        self.sgf.append(";{}[{}]\n".format(our_letter, sgf_move))
```

Letting your opponent play a move is structurally similar to your move. You ask GNU Go or Pachi to genmove a move, and you have to take care of converting the GTP response into a move that your bot understands. The only other thing you have to do is stop the game when your opponent resigns or both players pass.

Listing 8.24   Your opponent plays moves by responding to `genmove`

```
def play_their_move(self):
    their_name = self.their_color.name
    their_letter = their_name[0].upper()

    pos = self.command_and_response("genmove {}\n".format(their_name))
    if pos.lower() == 'resign':
        self.game_state = self.game_state.apply_move(Move.resign())
        self._stopped = True
    elif pos.lower() == 'pass':
        self.game_state = self.game_state.apply_move(Move.pass_turn())
        self.sgf.append(";{}[]\n".format(their_letter))
        if self.game_state.last_move.is_pass:
            self._stopped = True
    else:
        move = gtp_position_to_coords(pos)
        self.game_state = self.game_state.apply_move(move)
        self.sgf.append(";{}[{}]\n".format(their_letter,
            self.sgf.coordinates(move)))
```

That concludes your play_local.py implementation, and you can now test it as follows.

Listing 8.25   Letting one of your bots loose on Pachi

```
from dlgo.gtp.play_local import LocalGtpBot
from dlgo.agent.termination import PassWhenOpponentPasses
from dlgo.agent.predict import load_prediction_agent
import h5py

bot = load_prediction_agent(h5py.File("../agents/betago.hdf5", "r"))

gtp_bot = LocalGtpBot(go_bot=bot, termination=PassWhenOpponentPasses(),
                      handicap=0, opponent='pachi')
gtp_bot.run()
```

You should see the way the game between the bots unfolds, as shown in figure 8.4.

In the top part of the figure, you see the board printed by you, followed by your current estimate. In the lower half, you see Pachi's game state (which is identical to yours) on the left, and on the right Pachi gives you an estimation of its current assessment of the game in terms of which part of the board it thinks belongs to which player.

This is a hopefully convincing and exciting demo of what your bot can do by now, but it's not the end of the story. In the next section, we go one step further and show you how to connect your bot to a real-life Go server.

```
19  .  .  .  .  .  .  .  .  .  .  .  .  .  .  .  .  .  .  .
18  .  x  x  x  .  o  .  .  .  .  .  .  .  .  o  x  .  .  .
17  .  o  x  .  x  o  .  .  .  .  .  .  .  .  o  x  .  .  .
16  .  o  o  x  o  .  .  .  o  .  .  .  .  .  o  x  .  .  .
15  .  .  .  x  .  .  .  .  .  .  .  .  .  .  x  .  .  .  .
14  .  o  .  x  .  .  .  .  .  .  .  .  .  .  .  .  .  .  .
13  .  o  x  .  .  .  .  .  .  .  .  .  .  .  .  .  .  .  .
12  .  o  x  .  .  .  .  .  .  .  .  .  .  .  .  .  .  .  .
11  .  o  x  .  .  .  .  .  .  .  .  .  .  .  .  .  .  .  .
10  .  .  o  x  .  .  .  .  .  .  .  .  .  .  .  .  .  .  .
 9  .  .  o  x  .  .  .  .  .  .  .  .  .  .  .  .  x  .  .
 8  .  .  .  o  .  .  .  .  .  .  .  .  .  .  .  x  .  x  .
 7  .  .  .  .  .  .  .  o  o  .  .  .  .  .  o  o  x  .  .
 6  .  .  .  .  .  x  o  x  .  o  .  .  .  .  o  .  o  .  .
 5  .  .  x  x  .  x  .  x  .  .  .  .  .  .  o  .  o  .  .
 4  .  .  x  o  o  o  x  x  x  o  .  .  .  .  x  o  x  .  .
 3  .  x  o  .  .  x  o  x  .  .  o  .  .  .  x  .  x  .  .
 2  .  x  o  .  .  x  o  x  o  .  .  .  .  .  .  .  .  .  .
 1  .  .  .  .  .  .  .  .  .  .  .  .  .  .  .  .  .  .  .
    A  B  C  D  E  F  G  H  J  K  L  M  N  O  P  Q  R  S  T
Estimated result:
W+3.5
IN: genmove white
Move:  85  Komi: 7.5  Handicap: 0  Captures B: 4 W: 2
    A B C D E F G H J K L M N O P Q R S T        A B C D E F G H J K L M N O P Q R S T
    +------------------------------------+        +------------------------------------+
19 | . . . . . . . . . . . . . . . . . . . |   19 | x x x x , o o o o o o , , , , x x x x |
18 | . X X X . O . . . . . . . . O X . . . |   18 | , x x x , o o o o o , , , , o X x x x |
17 | . O X . X O . . . . . . . . O X . . . |   17 | o o X x x x o o o o o , , , , o X x x x |
16 | . O O X O . . . O . . . . . O X . . . |   16 | o o o X , , , o o o , , , , , X x x x |
15 | . . . X . . . . . . . . . . X ). . . |   15 | o o , X , , , , , , , , , , , , x x x |
14 | . O . X . . . . . . . . . . . . . . . |   14 | o o , X x , , , , , , , , , , , , , , |
13 | . O X . . . . . . . . . . . . . . . . |   13 | o o X X x , , , , , , , , , , , , , , |
12 | . O X . . . . . . . . . . . . . . . . |   12 | o o X X x , , , , , , , , , , , , , , |
11 | . O X . . . . . . . . . . . . . . . . |   11 | o o X X x , , , , , , , , , , , , , x |
10 | . . O X . . . . . . . . . . . . . . . |   10 | o o o X , , , , , , , , , , , , , x x |
 9 | . . O X . . . . . . . . . . . X . . |    9 | o o o X , , , , , , , , , , , , , x x |
 8 | . . . O . . . . . . . . . . . X . X |    8 | o o o o , , , , , , o , , , , , x x x |
 7 | . . . . . . . O O . . . . . O O X . |    7 | , , , , , , o o o o o , o o o x x x |
 6 | . . . . . X O X . O . . . . O . O . |    6 | , , , , x x X O X o o o o o o o o o , |
 5 | . . X X . X . X . . . . . . O . O . |    5 | x x x X X X X X X , o o o , , O o o o |
 4 | . . X O O O X X X O . . . . X O X . |    4 | X X X X X X X X X O o o , , , x o x , |
 3 | . X O . . X O X . . O . . . X . X . |    3 | X X X X X X X , , O o , , x , x , |
 2 | . X O . . X O X O . . . . . . . . . |    2 | X X X X X X X o o o o , , , , x x |
 1 | . . . . . . . . . . . . . . . . . . |    1 | X X X X X X X X , , o , , , , , x x |
    +------------------------------------+        +------------------------------------+
```

Figure 8.4 A snapshot of how Pachi and your bot see and evaluate a game between them

8.6   *Deploying a Go bot to an online Go server*

Note that play_local.py is really a tiny Go server for two bot opponents to play against each other. It accepts and sends GTP commands and knows when to start and finish a game. This produces overhead, because the program takes the role of a referee that controls how the opponents interact.

If you want to connect a bot to an actual Go server, this server will take care of all the game-play logic, and you can focus entirely on sending and receiving GTP commands. On the one hand, your fate becomes easier because you have less to worry about. On the other hand, connecting to a proper Go server means that you have to make sure to support the full range of GTP commands supported by that server, because otherwise your bot may crash.

To ensure that this doesn't happen, let's formalize the processing of GTP commands a little more. First, you implement a proper GTP response class for successful and failed commands.

Listing 8.26 Encoding and serializing a GTP response

```
class Response:
    def __init__(self, status, body):
        self.success = status
        self.body = body


def success(body=''):                        ⟵  Making a successful GTP
    return Response(status=True, body=body)       response with response body


def error(body=''):          ⟵  Making an error GTP response
    return Response(status=False, body=body)


def bool_response(boolean):        ⟵  Converting a Python Boolean into GTP
    return success('true') if boolean is True else success('false')


def serialize(gtp_command, gtp_response):    ⟵  Serializing a GTP response as a string
    return '{}{} {}\n\n'.format(
        '=' if gtp_response.success else '?',
        '' if gtp_command.sequence is None else str(gtp_command.sequence),
        gtp_response.body
    )
```

This leaves you with implementing the main class for this section, GTPFrontend. You put this class into frontend.py in the gtp module. You need the following imports, including command and response from your gtp module.

Listing 8.27 Python imports for your GTP frontend

```
import sys

from dlgo.gtp import command, response
from dlgo.gtp.board import gtp_position_to_coords, coords_to_gtp_position
from dlgo.goboard_fast import GameState, Move
from dlgo.agent.termination import TerminationAgent
from dlgo.utils import print_board
```

To initialize a GTP frontend, you need to specify an Agent instance and an optional termination strategy. GTPFrontend will then instantiate a dictionary of GTP events that you process. Each of these events, which includes common commands like play and others, will have to be implemented by you.

Listing 8.28 Initializing a `GTPFrontend`, which defines GTP event handlers

```
HANDICAP_STONES = {
    2: ['D4', 'Q16'],
    3: ['D4', 'Q16', 'D16'],
    4: ['D4', 'Q16', 'D16', 'Q4'],
    5: ['D4', 'Q16', 'D16', 'Q4', 'K10'],
    6: ['D4', 'Q16', 'D16', 'Q4', 'D10', 'Q10'],
    7: ['D4', 'Q16', 'D16', 'Q4', 'D10', 'Q10', 'K10'],
    8: ['D4', 'Q16', 'D16', 'Q4', 'D10', 'Q10', 'K4', 'K16'],
    9: ['D4', 'Q16', 'D16', 'Q4', 'D10', 'Q10', 'K4', 'K16', 'K10'],
}


class GTPFrontend:

    def __init__(self, termination_agent, termination=None):
        self.agent = termination_agent
        self.game_state = GameState.new_game(19)
        self._input = sys.stdin
        self._output = sys.stdout
        self._stopped = False

        self.handlers = {
            'boardsize': self.handle_boardsize,
            'clear_board': self.handle_clear_board,
            'fixed_handicap': self.handle_fixed_handicap,
            'genmove': self.handle_genmove,
            'known_command': self.handle_known_command,
            'komi': self.ignore,
            'showboard': self.handle_showboard,
            'time_settings': self.ignore,
            'time_left': self.ignore,
            'play': self.handle_play,
            'protocol_version': self.handle_protocol_version,
            'quit': self.handle_quit,
        }
```

After you start a game with the following run method, you continually read GTP commands that are forwarded to the respective event handler, which is done by the process method.

Listing 8.29 The frontend parses from the input stream until the game ends

```
    def run(self):
        while not self._stopped:
            input_line = self._input.readline().strip()
            cmd = command.parse(input_line)
            resp = self.process(cmd)
            self._output.write(response.serialize(cmd, resp))
            self._output.flush()

    def process(self, cmd):
        handler = self.handlers.get(cmd.name, self.handle_unknown)
        return handler(*cmd.args)
```

What's left to complete this GTPFrontend is the implementation of the individual GTP commands. The following listing shows the three most important ones; we refer you to the GitHub repository for the rest.

> Listing 8.30 A few of the most important event responses for your GTP frontend

```
def handle_play(self, color, move):
    if move.lower() == 'pass':
        self.game_state = self.game_state.apply_move(Move.pass_turn())
    elif move.lower() == 'resign':
        self.game_state = self.game_state.apply_move(Move.resign())
    else:
        self.game_state =
  self.game_state.apply_move(gtp_position_to_coords(move))
    return response.success()

def handle_genmove(self, color):
    move = self.agent.select_move(self.game_state)
    self.game_state = self.game_state.apply_move(move)
    if move.is_pass:
        return response.success('pass')
    if move.is_resign:
        return response.success('resign')
    return response.success(coords_to_gtp_position(move))

def handle_fixed_handicap(self, nstones):
    nstones = int(nstones)
    for stone in HANDICAP_STONES[nstones]:
        self.game_state = self.game_state.apply_move(
            gtp_position_to_coords(stone))
    return response.success()
```

You can now use this GTP frontend in a little script to start it from the command line.

> Listing 8.31 Starting your GTP interface from the command line

```
from dlgo.gtp import GTPFrontend
from dlgo.agent.predict import load_prediction_agent
from dlgo.agent import termination
import h5py

model_file = h5py.File("agents/betago.hdf5", "r")
agent = load_prediction_agent(model_file)
strategy = termination.get("opponent_passes")
termination_agent = termination.TerminationAgent(agent, strategy)

frontend = GTPFrontend(termination_agent)
frontend.run()
```

After this program runs, you can use it in exactly the same way you tested GNU Go in section 8.4: you can throw GTP commands at it, and it'll process them properly. Go ahead and test it by generating a move with genmove or printing out the board state with showboard. Any command covered in your event handler in GTPFrontend is feasible.

8.6.1 *Registering a bot at the Online Go Server*

Now that your GTP frontend is complete and works in the same way as GNU Go and Pachi locally, you can register your bots at an online platform that uses GTP for communication. You'll find that most popular Go servers are based on GTP, and appendix C covers three of them explicitly. One of the most popular servers in Europe and North America is the Online Go Server (OGS). We've chosen OGS as the platform to show you how to run a bot, but you could do the same thing with most other platforms as well.

Because the registration process for your bot at OGS is somewhat involved and the piece of software that connects your bot to OGS is a tool written in JavaScript, we've put this part into appendix E. You can either read this appendix now and come back here, or skip it if you're not interested in running your own bot online. When you complete appendix E, you'll have learned the following skills:

- Creating two accounts at OGS, one for your bot and one for you to administer your bot account
- Connecting your bot to OGS from your local computer for testing purposes
- Deploying your bot on an AWS instance to connect to OGS for as long as you wish

This will allow you to enter a (ranked) game against your own creation online. Also, everyone with an OGS account can play your bot at this point, which can be motivating to see. On top of that, your bot could even enter tournaments hosted on OGS!

8.7 *Summary*

- By building a deep network into your agent framework, you can make it so your models can interact with their environment.
- Registering an agent in a web application, by building an HTTP frontend, you can play against your own bots through a graphical interface.
- Using a cloud provider like AWS, you can rent compute power on a GPU to efficiently run your deep-learning experiments.
- Deploying your web application on AWS, you can easily share your bot and let it play with others.
- By letting your bot emit and receive Go Text Protocol (GTP) commands, it can play against other Go programs locally in a standardized way.
- Building a GTP frontend for your bot is the most important stepping stone to registering it at an online Go platform.
- Deploying a bot in the cloud, you can let it enter regular games and tournaments at the Online Go Server (OGS), and play against it yourself at any time.

Learning by practice: reinforcement learning

I've probably read a dozen books on Go, all written by strong pros from China, Korea, and Japan. And yet I'm just an intermediate amateur player. Why haven't I reached the level of these legendary players? Have I forgotten their lessons? I don't think that's it; I can practically recite Toshiro Kageyama's *Lessons in the Fundamentals of Go* (Ishi Press, 1978) by heart. Maybe I just need to read more books....

I don't know the full recipe for becoming a top Go star, but I know at least one difference between me and Go professionals: practice. A Go player probably clocks in five or ten thousand games before qualifying as a professional. Practice creates knowledge, and sometimes that's knowledge that you can't directly communicate. You can *summarize* that knowledge—that's what makes it into Go books. But the subtleties get lost in the translation. If I expect to master the lessons I've read, I need to put in a similar level of practice.

If practice is so valuable for humans, what about computers? Can a computer program learn by practicing? That's the promise of *reinforcement learning*. In reinforcement learning (RL), you improve a program by having it repeatedly attempt a task. When it has good outcomes, you modify the program to repeat its decisions. When it has bad outcomes, you modify the program to avoid those decisions. This doesn't mean you write new code after each trial: RL algorithms provide automated methods for making those modifications.

Reinforcement learning isn't a free lunch. For one thing, it's slow: your bot will need to play thousands of games in order to make a measurable improvement. In addition, the training process is fiddly and hard to debug. But if you put in the effort to make these techniques work for you, the payoff is huge. You can build software that applies sophisticated strategies to tackle a variety of tasks, even if you can't describe those strategies yourself.

This chapter starts with a birds-eye view of the reinforcement-learning cycle. Next, you'll see how to set up a Go bot to play against itself in a way that fits into the reinforcement-learning process. Chapter 10 shows how to use the self-play data to improve your bot's performance.

9.1 The reinforcement-learning cycle

Many algorithms implement the mechanics of reinforcement learning, but they all work within a standard framework. This section describes the reinforcement-learning cycle, in which a computer program improves by repeatedly attempting a task. Figure 9.1 illustrates the cycle.

In the language of reinforcement learning, your Go bot is an *agent*: a program that makes decisions in order to accomplish a task. Earlier in the book, you implemented several versions of an `Agent` class that could choose Go moves. In those cases, you provided the agent with a situation—a `GameState` object—and it responded with a decision—a move to play. Although you weren't using reinforcement learning at that time, the concept of an agent is the same.

The goal of reinforcement learning is to make the agent as effective as possible. In this case, you want your agent to win at Go.

First, you have your Go bot play a batch of games against itself; during each game, it should record every turn and the final outcome. These game records are called its *experience*.

Next, you *train* your bot by updating its behavior in response to what happened in its self-play games. This process is similar to training the neural networks covered in chapters 6 and 7. The core idea is that you want the bot to repeat the decisions it made in games it won, and stop making the decisions it made in games it lost. The training algorithm comes as a package deal with the structure of your agent: you need to be able to systematically modify the behavior of your agent in order to train. There are many algorithms for doing this; we cover three in this book. In this chapter and the next, we

General reinforcement-learning cycle

Update the agent's behavior in
response to the experience results.

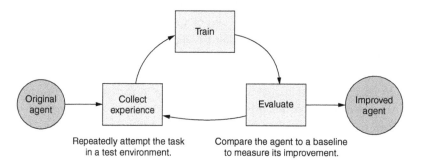

Repeatedly attempt the task   Compare the agent to a baseline
in a test environment.   to measure its improvement.

Reinforcement-learning cycle for Go AI

Update neural network
weights from the game
results by using gradient descent.

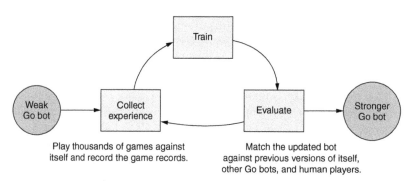

Play thousands of games against   Match the updated bot
itself and record the game records.   against previous versions of itself,
                                      other Go bots, and human players.

Figure 9.1   The reinforcement-learning cycle. You can implement reinforcement learning in many ways, but the overall process has a common structure. First, a computer program attempts a task repeatedly. The records of these attempts are called *experience data*. Next, you modify the behavior to imitate the more successful attempts; this process is *training*. You then periodically evaluate the performance to confirm that the program is improving. Normally, you need to repeat this process for many cycles.

start with the *policy gradient* algorithm. In chapter 11, we cover the *Q-learning* algorithm. Chapter 12 introduces the *actor-critic* algorithm.

After training, you expect your bot to be a bit stronger. But there are many ways for the training process to go wrong, so it's a good idea to evaluate the bot's progress to confirm its strength. To evaluate a game-playing agent, have it play more games. You can pit your agent against earlier versions of itself to measure its progress. As a sanity check, you can also periodically compare your bot to other AIs or play against it yourself.

Then you can repeat this entire cycle indefinitely:

- Collect experience
- Train
- Evaluate

We'll break this cycle into multiple scripts. In this chapter, you'll implement a self_play script that will simulate the self-play games and save the experience data to disk. In the next chapter, you'll make a train script that takes the experience data as input, updates the agent accordingly, and saves the new agent.

9.2   *What goes into experience?*

In chapter 3, you designed a set of data structures for representing Go games. You can imagine how you could store an entire game record by using classes such as Move, GoBoard, and GameState. But reinforcement-learning algorithms are generic: they deal with a highly abstract representation of a problem, so that the same algorithms can apply to as many problem domains as possible. This section shows how to describe game records in the language of reinforcement learning.

In the case of game playing, you can divide your experience into individual games, or *episodes*. An episode has a clear end, and decisions made during one episode have no bearing on what happens in the next. In other domains, you may not have any obvious way to divide the experience into episodes; for example, a robot that's designed to operate continuously makes an endless sequence of decisions. You can still apply reinforcement learning to such problems, but the episode boundaries here make it a little simpler.

Within an episode, an agent is faced with a *state* of its environment. Based on the current state, the agent must select an *action*. After choosing an action, the agent sees a new state; the next state depends on both the chosen action and whatever else is going on in the environment. In the case of Go, your AI will see a board position (the state), and then select a legal move (an action). After that, the AI will see a new board position on its next turn (the next state).

Note that after the agent chooses an action, the next state also includes the opponent's move. You can't determine the next state from the current state and the action you choose: you must also wait for the opponent's move. The opponent's behavior is part of the *environment* that your agent must learn to navigate.

In order to improve, your agent needs feedback about whether it's achieving its objective. You provide that feedback by calculating its *reward*, a numerical score for meeting a goal. For your Go AI, the goal is to win a game, so you'll communicate a reward of 1 each time it wins and –1 each time it loses. Reinforcement-learning algorithms will modify the agent's behavior so as to increase the amount of reward it accumulates. Figure 9.2 illustrates how a game of Go can be described with states, actions, and rewards.

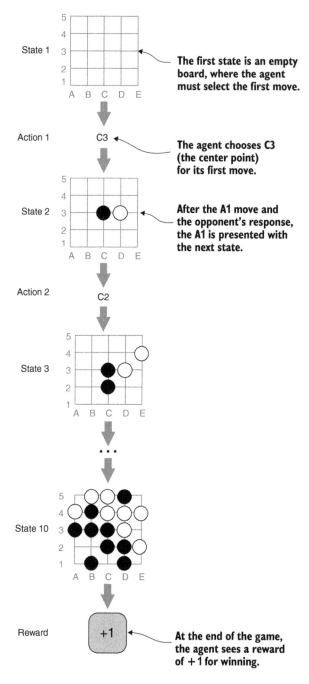

Figure 9.2   A game of 5 × 5 Go translated into the language of reinforcement learning. The agent that you want to train is the black player. It sees a sequence of states (board positions) and chooses actions (legal moves). At the end of an episode (a complete game), it gets a reward to indicate whether it achieved its goal. In this case, black wins the game, so the agent sees a reward of +1.

Go and similar games are special cases: the reward comes all at once, at the end of the game. And there are only two possible rewards: you win or you lose, and you don't care about what else happens in the game. In other domains, the reward may be spread out. Imagine making an AI to play Scrabble. On each turn, the AI will place a word and score points, and then its opponent will do the same. In that case, you can compute a positive reward for the AI's points, and a negative reward for the opponent's points. Then the AI doesn't have to wait all the way to the end of an episode for its reward; it gets little pieces of its reward after every action it takes.

A key idea in reinforcement learning is that an action may be responsible for a reward that comes much later. Imagine you make an especially clever play on move 35 of a game, and continue on to win after 200 moves. Your good move early on deserves at least some of the credit for the win. You must somehow split up the credit for the reward over all the moves in the game. The future reward that your agent sees after an action is called the *return* on that action. To compute the return on an action, you add up all the rewards the agent saw after that action, all the way to the end of the episode, as shown in listing 9.1. This is a way of saying that you don't know, in advance, which moves are responsible for winning or losing. The onus is on the learning algorithm to split up the credit or blame among individual moves.

Listing 9.1   Calculating return on an action

reward[i] is the reward the agent saw immediately after action i.

```
for exp_idx in range(exp_length):
    total_return[exp_idx] = reward[exp_idx]
    for future_reward_idx in range(exp_idx + 1, exp_length):
        total_return[exp_idx] += reward[future_reward_idx]
```

Loops over all future rewards and adds them into the return

That assumption doesn't make sense for every problem. Consider our Scrabble example again. The decisions you make on your first turn could plausibly affect your score on your third turn—maybe you held a high-scoring X in reserve until you could combine it with a bonus square. But it's hard to see how decisions on your third turn could affect your twentieth. To represent this concept in your return calculation, you can compute a weighted sum of the future rewards from each action. The weights should get smaller as you go further from the action, so that far-future rewards have less influence than immediate rewards.

This technique is called *discounting* the reward. Listing 9.2 shows how to calculate discounted returns. In that example, each action gets full credit for the reward that comes immediately after. But the reward from the next step counts for only 75% as much; the reward two steps out counts 75% × 75% = 56% as much; and so on. The choice of 75% is just an example; the correct discount rate will depend on your particular domain, and you may need to experiment a bit to find the most effective number.

Listing 9.2   Calculating discounted returns

```
for exp_idx in range(exp_length):
    discounted_return[exp_idx] = reward[exp_idx]
    discount_amount = 0.75
    for future_reward_idx in range(exp_idx + 1, exp_length):
        discounted_return[exp_idx] +=
            discount_amount * reward[future_reward_idx]
        discount_amount *= 0.75
```

The discount _amount gets smaller and smaller as you get further from the original action.

In the case of building a Go AI, the only possible reward is a win or loss. This lets you take a shortcut in the return calculation. When your agent wins, every action in the game has a return of 1. When your agent loses, every action has a return of –1.

9.3   *Building an agent that can learn*

Reinforcement learning can't create a Go AI, or any other kind of agent, out of thin air. It can only *improve* a bot that already works within the parameters of the game. To get started, you need an agent that can at least complete a game. This section shows how to create a Go bot that selects moves by using a neural network. If you start with an untrained network, the bot will play as badly as your original RandomAgent from chapter 3. Later, you can improve this neural network through reinforcement learning.

A *policy* is a function that selects an action from a given state. In earlier chapters, you saw several implementations of the Agent class that have a select_move function. Each of those select_move functions is a policy: a game state comes in, and a move comes out. All the policies you've implemented so far are valid, in the sense that they produce legal moves. But they're not equally good: the MCTSAgent from chapter 4 will defeat the RandomAgent from chapter 3 more often than not. If you want to improve one of these agents, you need to think of an improvement to the algorithm, write new code, and test it—the standard software development process.

To use reinforcement learning, you need a policy that you can update automatically, using another computer program. In chapter 6, you studied a class of functions that lets you do exactly that: convolutional neural networks. A deep neural network can compute sophisticated logic, and you can modify its behavior by using the gradient descent algorithm.

The move-prediction neural network you designed in chapters 6 and 7 outputs a vector with a value for each point on the board; the value represents the network's confidence that point would be the next play. How can you form a policy from such an output? One way is to simply select the move with the highest value. This will produce good results if your network has already been trained to select good moves. But it'll always select the same move for any given board position. This creates a problem for reinforcement learning. To improve through reinforcement learning, you need to select a variety of moves. Some will be better, and some will be worse; you can detect the good moves by looking at the outcomes they produce. But you need the variety in order to improve.

Instead of always selecting the highest-rated move, you want a stochastic policy. Here, *stochastic* means that if you input the exact same board position twice, your agent may select different moves. This involves randomness, but not in the same way as your RandomAgent from chapter 3. The RandomAgent chose moves with no regard to what was happening in the game. A stochastic policy means that your move selection will depend on the state of the board, but it won't be 100% predictable.

9.3.1 *Sampling from a probability distribution*

For any board position, your neural network will give you a vector with one element for each board position. To create a policy from this, you can treat each element of the vector as indicating the probability that you select a particular move. This section shows how to select moves according to those probabilities.

For example, if you're playing rock-paper-scissors, you could follow a policy of choosing rock 50% of the time, paper 30% of the time, and scissors 20% of the time. The 50%-30%-20% split is a *probability distribution* over the three choices. Note that probabilities sum to exactly 100%: this is because your policy must always choose exactly one item from the list. This is a necessary property of a probability distribution; a 50%-30%-10% policy would leave you with no decision 10% of the time.

The process of randomly selecting one of those items in those proportions is called *sampling* from that probability distribution. The following listing shows a Python function that will choose one of those options according to that policy.

> **Listing 9.3   An example of sampling from a probability distribution**

```
import random

def rps():
    randval = random.random()
    if 0.0 <= randval < 0.5:
        return 'rock'
    elif 0.5 <= randval < 0.8:
        return 'paper'
    else:
        return 'scissors'
```

Try this snippet out a few times and see how it behaves. You'll see rock more than paper, and paper more than scissors. But all three will appear regularly.

This logic for sampling from a probability distribution is built into NumPy as the np.random.choice function. The following listing shows the exact same behavior implemented with NumPy.

> **Listing 9.4   Sampling from a probability distribution with NumPy**

```
import numpy as np

def rps():
```

```
    return np.random.choice(
        ['rock', 'paper', 'scissors'],
        p=[0.5, 0.3, 0.2])
```

In addition, np.random.choice will handle *repeated* sampling from the same distribution. It'll sample from your distribution once, remove that item from the list, and sample again from the remaining items. In this way, you get a semirandom ordered list. The high-probability items are likely to appear near the front of the list, but some variety remains. The following listing shows how to get repeated sampling with np.random.choice. You pass size=3 to indicate that you want three different items, and replace=False to indicate that you don't want any results repeated.

> **Listing 9.5   Repeatedly sampling from a probability distribution with NumPy**

```
import numpy as np

def repeated_rps():
    return np.random.choice(
        ['rock', 'paper', 'scissors'],
        size=3,
        replace=False,
        p=[0.5, 0.3, 0.2])
```

The repeated sampling will be useful in case your Go policy recommends an invalid move. In that case, you'll want to select another one. You can call np.random.choice once and then just work your way down the list it generates.

9.3.2 Clipping a probability distribution

The reinforcement-learning process can be fairly unstable, especially early on. The agent may overreact to a few chance wins and temporarily assign a high probability to moves that really aren't that good. (In that respect, it's not unlike human beginners!) It's possible for the probability for a particular move to go all the way to 1. This creates a subtle problem: because your agent will always select the same move, it has no opportunity to unlearn it.

To prevent this, you'll *clip* the probability distribution to make sure no probabilities get pushed all the way to 0 or 1. You did the same with the DeepLearningAgent from chapter 8. The np.clip function from NumPy handles most of the work here.

> **Listing 9.6   Clipping a probability distribution**

```
def clip_probs(original_probs):
    min_p = 1e-5
    max_p = 1 - min_p
    clipped_probs = np.clip(original_probs, min_p, max_p)
    clipped_probs = clipped_probs / np.sum(clipped_probs)    ◁──┐  Ensure that the
    return clipped_probs                                          result is still a
                                                                  valid probability
                                                                  distribution.
```

9.3.3 Initializing an agent

Let's start building out a new type of agent, a `PolicyAgent`, that selects moves according to a stochastic policy and can learn from experience data. This model can be identical to the move-prediction model from chapters 6 and 7; the only difference is in how you train it. You'll add this to your dlgo library in the dlgo/agent/pg.py module.

Recall from the previous chapters that your model needs a matching board-encoding scheme. The `PolicyAgent` class can accept the model and board encoder in the constructor. This creates a nice separation of concerns. The `PolicyAgent` class is responsible for selecting moves according to the model and changing its behavior in response to its experience. But it can ignore the details of the model structure and the board-encoding scheme.

Listing 9.7   The constructor for the `PolicyAgent` class

```
class PolicyAgent(Agent):
    def __init__(self, model, encoder):
        self.model = model
        self.encoder = encoder
```

Implements the Encoder interface

A Keras Sequential model instance

To start the reinforcement-learning process, you first construct a board encoder, then a model, and finally the agent. The following listing shows this process.

Listing 9.8   Constructing a new learning agent

```
encoder = encoders.simple.SimpleEncoder((board_size, board_size))
model = Sequential()
for layer in dlgo.networks.large.layers(encoder.shape()):
    model.add(layer)
model.add(Dense(encoder.num_points()))
model.add(Activation('softmax'))
new_agent = agent.PolicyAgent(model, encoder)
```

Adds an output layer that will return a probability distribution over points on the board

Builds a Sequential model out of the layers described in dlgo.networks.large (covered in chapter 6)

When you construct an agent like this, using a newly created model, Keras initializes the model weights to small, random values. At this point, the agent's policy will be close to *uniform random*: it'll choose any valid move with roughly equal probability. Later, training the model will add structure to its decisions.

9.3.4 Loading and saving your agent from disk

The reinforcement-learning process can continue indefinitely; you may spend days or even weeks training your bot. You'll want to periodically persist your bot to disk so you can start and stop the training process, and compare its performance at different points in the training cycle.

You can use the HDF5 file format, which we introduced in chapter 8, to store your agent. The HDF5 format is a convenient way to store numerical arrays, and it integrates nicely with NumPy and Keras.

A `serialize` method on your `PolicyAgent` class can persist its encoder and model to disk, which is enough to re-create the agent.

Listing 9.9 Serializing a `PolicyAgent` to disk

```
class PolicyAgent(Agent):                     Stores enough information to
...                                           reconstruct the board encoder
    def serialize(self, h5file):
        h5file.create_group('encoder')
        h5file['encoder'].attrs['name'] = self.encoder.name()
        h5file['encoder'].attrs['board_width'] = \
            self.encoder.board_width
        h5file['encoder'].attrs['board_height'] = \
            self.encoder.board_height
        h5file.create_group('model')
        kerasutil.save_model_to_hdf5_group(       Uses built-in Keras features to
            self._model, h5file['model'])         persist the model and its weights
```

The `h5file` argument could be an `h5py.File` object, or it could be a group inside an `h5py.File`. This allows you to bundle other data with the agent in a single HDF5 file.

To use this `serialize` method, you first create a new HDF5 file, and then pass in the file handle.

Listing 9.10 An example of using the `serialize` function

```
import h5py

with h5py.File(output_file, 'w') as outf:
    agent.serialize(outf)
```

Then a corresponding `load_policy_agent` function reverses the procedure.

Listing 9.11 Loading a policy agent from a file

```
                                              Uses built-in Keras functions
                                              to load the model structure
def load_policy_agent(h5file):                and weights
    model = kerasutil.load_model_from_hdf5_group(
        h5file['model'])
    encoder_name = h5file['encoder'].attrs['name']
    board_width = h5file['encoder'].attrs['board_width']
    board_height = h5file['encoder'].attrs['board_height']    Recovers the
    encoder = encoders.get_encoder_by_name(                   board encoder
        encoder_name,
        (board_width, board_height))
    return PolicyAgent(model, encoder)        Reconstructs the agent
```

9.3.5 *Implementing move selection*

The `PolicyAgent` needs one more function before you can begin self-play: the `select_move` implementation. This function will look similar to the `select_move` function you added to the `DeepLearningAgent` from chapter 8. The first step is to encode the board as a tensor (a stack of matrices; see appendix A) suitable for feeding into the model. Next, you feed the board tensor to the model and get back a probability distribution of the moves. You then clip the distribution to make sure no probability goes all the way to 1 or 0. Figure 9.3 illustrates the flow of this process. Listing 9.12 shows how to implement these steps.

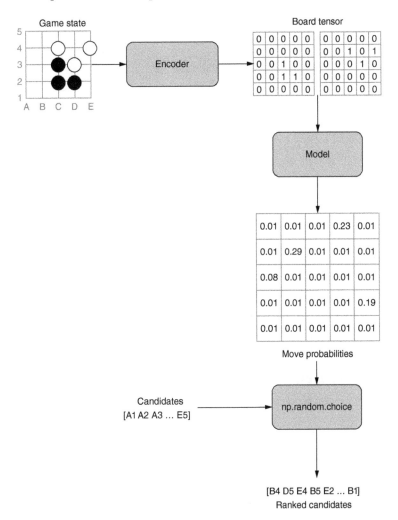

Figure 9.3 The move-selection process. First you encode a game state as a numerical tensor; then you can pass that tensor to your model to get move probabilities. You sample from all points on the board according to the move probabilities to get an order in which to try the moves.

Listing 9.12 **Selecting a move with a neural network**

```
class PolicyAgent(Agent):
...
    def select_move(self, game_state):
        board_tensor = self._encoder.encode(game_state)
        X = np.array([board_tensor])
        move_probs = self._model.predict(X)[0]

        move_probs = clip_probs(move_probs)

        num_moves = self._encoder.board_width * \
            self._encoder.board_height
        candidates = np.arange(num_moves)
        ranked_moves = np.random.choice(
            candidates, num_moves,
            replace=False, p=move_probs)

        for point_idx in ranked_moves:
            point = self._encoder.decode_point_index(point_idx)
            move = goboard.Move.play(point)
            is_valid = game_state.is_valid_move(move)
            is_an_eye = is_point_an_eye(
                game_state.board,
                point,
                game_state.next_player)
            if is_valid and (not is_an_eye):
                return goboard.Move.play(point)
        return goboard.Move.pass_turn()
```

The Keras predict call makes batch predictions, so you wrap your single board in an array and pull out the first item from the resulting array.

Creates an array containing the index of every point on the board

Samples from the points on the board according to the policy, creates a ranked list of points to try

Loops over each point, checks if it's a valid move, and picks the first valid one

If you fall through here, there are no reasonable moves left.

9.4 Self-play: how a computer program practices

Now that you have a learning agent capable of completing a game, you can begin collecting experience data. For a Go AI, this means playing thousands of games. This section shows how to implement this process. First, we describe some data structures to make handling experience data more convenient. Next, we show how to implement the self-play driver program.

9.4.1 Representing experience data

Experience data contains three parts: states, actions, and rewards. To help keep these organized, you can create a single data structure that holds all three of these together.

The ExperienceBuffer class is a minimal container for an experience data set. It has three attributes: states, actions, and rewards. All of these are represented as NumPy arrays; your agent will be responsible for encoding its states and actions as numerical structures. The ExperienceBuffer is nothing more than a container for passing the data set around. Nothing in this implementation is specific to policy gradient learning; you can reuse this class with other RL algorithms in later chapters. So you'll add this class to the dlgo/rl/experience.py module.

Listing 9.13 Constructor for an experience buffer

```
class ExperienceBuffer:
    def __init__(self, states, actions, rewards):
        self.states = states
        self.actions = actions
        self.rewards = rewards
```

After you've collected a large experience buffer, you'll want a way to persist it to disk. The HDF5 file format is a perfect fit once again. You can add a `serialize` method to the `ExperienceBuffer` class.

Listing 9.14 Saving an experience buffer to disk

```
class ExperienceBuffer:
...
    def serialize(self, h5file):
        h5file.create_group('experience')
        h5file['experience'].create_dataset(
            'states', data=self.states)
        h5file['experience'].create_dataset(
            'actions', data=self.actions)
        h5file['experience'].create_dataset(
            'rewards', data=self.rewards)
```

You'll also need a corresponding function, `load_experience`, to read the experience buffer back out of the file. Note that you cast each data set to `np.array` when reading it: that'll read the entire dataset into memory.

Listing 9.15 Restoring an `ExperienceBuffer` from an HDF5 file

```
def load_experience(h5file):
    return ExperienceBuffer(
        states=np.array(h5file['experience']['states']),
        actions=np.array(h5file['experience']['actions']),
        rewards=np.array(h5file['experience']['rewards']))
```

Now you have a simple container for passing around experience data. You still need a way to fill it with your agent's decisions. The complication is that the agent makes decisions one at a time, but it doesn't get a reward until the game is over and you know who won. To resolve this, you need to keep track of all the decisions from the current episode until it's complete. One option is to put this logic directly in the agent, but this will clutter up the implementation of `PolicyAgent`. Alternately, you can separate this out into a discrete `ExperienceCollector` object whose sole responsibility is episode-by-episode bookkeeping.

The `ExperienceCollector` implements four methods:

- `begin_episode` and `complete_episode`, which are called by the self-play driver to indicate the start and end of a single game.

- record_decision, which is called by the agent to indicate a single action it chose.
- to_buffer, which packages up everything the ExperienceCollector has recorded and returns an ExperienceBuffer. The self-play driver will call this at the end of a self-play session.

The full implementation appears in the following listing.

Listing 9.16   An object to track decisions within a single episode

```
class ExperienceCollector:
    def __init__(self):
        self.states = []
        self.actions = []
        self.rewards = []
        self.current_episode_states = []
        self.current_episode_actions = []

    def begin_episode(self):
        self.current_episode_states = []
        self.current_episode_actions = []

    def record_decision(self, state, action):
        self.current_episode_states.append(state)
        self.current_episode_actions.append(action)

    def complete_episode(self, reward):
        num_states = len(self.current_episode_states)
        self.states += self.current_episode_states
        self.actions += self.current_episode_actions
        self.rewards += [reward for _ in range(num_states)]

        self.current_episode_states = []
        self.current_episode_actions = []

    def to_buffer(self):
        return ExperienceBuffer(
            states=np.array(self.states),
            actions=np.array(self.actions),
            rewards=np.array(self.rewards)
        )
```

> Saves a single decision in the current episode; the agent is responsible for encoding the state and action.

> Spreads the final reward across every action in the game

> The ExperienceCollector accumulates Python lists; this converts them to NumPy arrays.

To integrate the ExperienceCollector with your agent, you can add a set_collector method that tells the agent where to send its experiences. Then inside select_move, the agent will notify the collector every time it makes a decision.

Listing 9.17   Integrating an ExperienceCollector with a PolicyAgent

```
class PolicyAgent:
...
    def set_collector(self, collector):
        self.collector = collector
```

> Allows the self-play driver program to attach a collector to the agent

```
...
    def select_move(self, game_state):
...
        if self.collector is not None:
                self.collector.record_decision(
                    state=board_tensor,
                    action=point_idx
                )
        return goboard.Move.play(point)
```

> **At the time it chooses a move, notifies the collector of the decision**

9.4.2 Simulating games

The next step is playing the games. You've done this twice before in the book: in the bot_v_bot demo in chapter 3, and as part of the Monte Carlo tree-search implementation in chapter 4. You can use the same simulate_game implementation here.

Listing 9.18 Simulating a game between two agents

```
def simulate_game(black_player, white_player):
    game = GameState.new_game(BOARD_SIZE)
    agents = {
        Player.black: black_player,
        Player.white: white_player,
    }
    while not game.is_over():
        next_move = agents[game.next_player].select_move(game)
        game = game.apply_move(next_move)
    game_result = scoring.compute_game_result(game)
    return game_result.winner
```

In this function, black_player and white_player could be any instance of your Agent class. You can match up the PolicyAgent that you're training against any opponent you like. Theoretically, the opponent could be a human player, although it'd take ages to collect enough experience data that way. Or your learner could play against a third-party Go bot, perhaps using the GTP framework from chapter 8 to handle the communications.

You can also just match up your learning agent with a copy of itself. Besides the simplicity of this solution, there are two specific advantages.

First, reinforcement learning needs plenty of both successes and failures to learn from. Imagine playing your first-ever game of chess or Go against a grandmaster. As a novice, you'd be so far behind it would be impossible to tell where you went wrong, and the experienced player could probably make a few mistakes and still win comfortably. As a result, neither player would learn much from the game. Instead, beginners usually start against other beginners and work their way up slowly. The same principle applies in reinforcement learning. When your bot plays itself, it'll always have an equal-strength opponent.

Second, by playing your agent against itself, you get two games for the price of one. Because the same decision-making process went into both sides of the game, you can learn from both the winning side and the losing side. You'll need huge volumes of games for reinforcement learning, so generating them twice as fast is a nice bonus.

To start the self-play process, you construct two copies of your agent and assign them each an `ExperienceCollector`. Each agent needs its own collector because the two agents will see different rewards at the end of a game. Listing 9.19 shows this initialization step.

Reinforcement learning beyond games

Self-play is a great technique for collecting experience data for board games. In other domains, you'll need to separately build a simulated environment to run your agent. For example, if you want to use reinforcement learning to build a control system for a robot, you'd need a detailed simulation of the physical environment the robot will operate in.

If you want to experiment further with reinforcement learning, the OpenAI Gym (https://github.com/openai/gym) is a useful resource. It provides environments for a variety of board games, video games, and physical simulations.

Listing 9.19   Initialization for generating a batch of experience

```
agent1 = agent.load_policy_agent(h5py.File(agent_filename))
agent2 = agent.load_policy_agent(h5py.File(agent_filename))
collector1 = rl.ExperienceCollector()
collector2 = rl.ExperienceCollector()
agent1.set_collector(collector1)
agent2.set_collector(collector2)
```

Now you're ready to implement the main loop that simulates the self-play games. In this loop, `agent1` will always play as black, while `agent2` will always play as white. This is fine, so long as `agent1` and `agent2` are identical and you intend to combine their experiences for training. If your learning agent is playing against another reference agent, you'll want it to alternate between black and white. In Go, black and white have slightly different personalities due to black playing first, so a learning agent needs to practice from both sides.

Listing 9.20   Playing a batch of games

```
for i in range(num_games):
    collector1.begin_episode()
    collector2.begin_episode()

    game_record = simulate_game(agent1, agent2)
    if game_record.winner == Player.black:
        collector1.complete_episode(reward=1)        agent1 won the game, so it
        collector2.complete_episode(reward=-1)       gets a positive reward.
    else:
```

```
collector2.complete_episode(reward=1)        agent2 won the game.
collector1.complete_episode(reward=-1)
```

When the self-play is complete, the last step is to combine all the collected experience and save it in a file. That file provides the input for the training script, which we cover in the next chapter.

Listing 9.21 Saving a batch of experience data

```
experience = rl.combine_experience([        ◁──┐ Merges both agents' experience
    collector1,                                │ into a single buffer
    collector2])
with h5py.File(experience_filename, 'w') as experience_outf:   ◁──┐ Saves into
    experience.serialize(experience_outf)                          │ an HDF5 file
```

At this point, you're ready to generate self-play games. The next chapter shows you how to start improving your bot from the self-play data.

9.5 Summary

- An *agent* is a computer program that's supposed to accomplish a certain task. For example, our Go-playing AI is an agent with the goal of winning games of Go.
- The reinforcement-learning cycle involves collecting experience data, training the agent from the experience data, and evaluating the updated agent. At the end of a cycle, you expect a small improvement in your agent's performance. Ideally, you can repeat this cycle many times to continually improve your agent.
- To apply reinforcement learning to a problem, you must describe the problem in terms of *states*, *actions*, and *rewards*.
- Rewards are the way you control the behavior of your reinforcement-learning agent. You can provide positive rewards for outcomes you want your agent to achieve, and negative rewards for outcomes you want your agent to avoid.
- A *policy* is a rule for making decisions from a given state. In a Go AI, the algorithm that selects a move from a board position is its policy.
- You can make a policy out of a neural network by treating the output vector as a *probability distribution* over possible actions, and then *sampling* from the probability distribution.
- When applying reinforcement learning to games, you can collect experience data through *self-play*: your agent plays games against a copy of itself.

Reinforcement learning with policy gradients

This chapter covers

- Improving game play with policy gradient learning
- Implementing policy gradient learning in Keras
- Tuning optimizers for policy gradient learning

Chapter 9 showed you how to make a Go-playing program play against itself and save the results in experience data. That's the first half of reinforcement learning; the next step is to use experience data to improve the agent so that it wins more often. The agent from the previous chapter used a neural network to select which move to play. As a thought experiment, imagine you shift every weight in the network by a random amount. Then the agent will select different moves. Just by luck, some of those new moves will be better than the old ones; others will be worse. On balance, the updated agent might be slightly stronger or weaker than the previous version. Which way it goes is up to chance.

Can you improve on that? This chapter covers a form of *policy gradient learning*. Policy gradient methods provide a scheme for estimating which direction to shift the weights in order to make the agent better at its task. Instead of randomly shifting each weight, you can analyze the experience data to guess whether it's better to increase or decrease a particular weight. Randomness still plays a role, but policy gradient learning improves your odds.

Recall from chapter 9 that you're making decisions with a stochastic policy—a function that specifies a probability for each possible move the agent can make. The policy-learning method we cover in this chapter works like this:

1   When the agent wins, increase the probability of each move it picked.
2   When the agent loses, decrease the probability of each move it picked.

First, you'll work through a simplified example to show how improving a policy by this technique can lead to winning more games. Next, you'll see how to use gradient descent to make the change you want—increasing or decreasing the probability of a specific move—in a neural network. We wrap up with some practical tips for managing the training process.

10.1 How random games can identify good decisions

To introduce policy learning, we'll start with a game that's much simpler than Go. Let's call this game Add It Up. Here are the rules:

- On each turn, each player picks a number between 1 and 5.
- After 100 turns, each player adds up all the numbers they chose.
- The player with the higher total wins.

Yes, this means the optimal strategy is to just pick 5 on each turn. No, this isn't a good game. We'll use this game to illustrate *policy learning*, where you gradually improve a stochastic policy based on game results. Because you know the correct strategy for this game, you can see how policy learning leads toward perfect play.

Add It Up is a shallow game, but we can use it as a metaphor for a more serious game like Go. Just as in Go, a game of Add It Up is long, and players have many opportunities to make good plays or blunders within the same game. To update a policy from game results, you need to identify which moves deserve credit or blame for winning or losing a particular game. This is called the *credit assignment* problem, and it's one of the core problems in reinforcement learning. This section demonstrates how you can average many game results to assign credit to individual decisions. In chapter 12, you'll build on this technique to create a more sophisticated and robust credit assignment algorithm.

Let's start with a purely random policy, where you select any of the five options with equal probability. (Such a policy is called *uniform random*.) Over the course of a full game, you'd expect that policy to select 1 about 20 times, 2 about 20 times, 3 about 20 times, and so on. But you won't expect 1 to appear *exactly* 20 times; it'll vary from game to game. The following listing shows a Python function that simulates all the choices such an agent will make over a game. Figure 10.1 shows the results of a few sample runs; you can try the snippet yourself a few times and see.

```
import numpy as np

counts = {1: 0, 2: 0, 3: 0, 4: 0, 5: 0}
for i in range(100):
    choice = np.random.choice([1, 2, 3, 4, 5],
                              p=[0.2, 0.2, 0.2, 0.2, 0.2])
    counts[choice] += 1
print(counts)
```

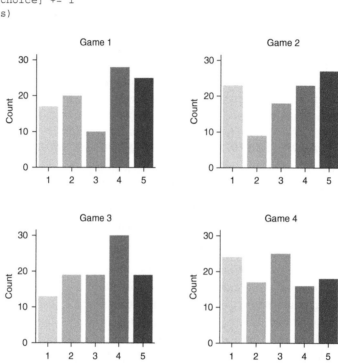

Figure 10.1   This graph shows four sample games played by a random agent. The bars indicate how often the agent chose each of the five possible moves in each game. Although the agent used the same policy in all the games, the exact counts vary quite a bit from game to game.

Even though the agent follows the exact same policy every game, the stochastic nature of the policy causes variance from game to game. You can exploit that variance to improve the policy.

The following listing shows a function that simulates a complete game of Add It Up, tracks the decisions each player makes, and calculates the winner.

```
def simulate_game(policy):
    """Returns a tuple of (winning choices, losing choices)"""
    player_1_choices = {1: 0, 2: 0, 3: 0, 4: 0, 5: 0}
    player_1_total = 0
```

```
player_2_choices = {1: 0, 2: 0, 3: 0, 4: 0, 5: 0}
player_2_total = 0
for i in range(100):
    player_1_choice = np.random.choice([1, 2, 3, 4, 5],
                                       p=policy)
    player_1_choices[player_1_choice] += 1
    player_1_total += player_1_choice
    player_2_choice = np.random.choice([1, 2, 3, 4, 5],
                                       p=policy)
    player_2_choices[player_2_choice] += 1
    player_2_total += player_2_choice
if player_1_total > player_2_total:
    winner_choices = player_1_choices
    loser_choices = player_2_choices
else:
    winner_choices = player_2_choices
    loser_choices = player_1_choices
return (winner_choices, loser_choices)
```

Run a few games and look at the results; listing 10.3 shows some example runs. Usually, the winner picks 1 less often, but not always. Sometimes the winner picks 5 more often, but that's not guaranteed either.

Listing 10.3   Sample outputs of listing 10.2

```
>>> policy = [0.2, 0.2, 0.2, 0.2, 0.2]
>>> simulate_game(policy)
({1: 20, 2: 23, 3: 15, 4: 25, 5: 17},        ⟵┘ Winner's choices
 {1: 21, 2: 20, 3: 24, 4: 16, 5: 19})        ⟵┐ Loser's choices
>>> simulate_game(policy)
({1: 22, 2: 22, 3: 19, 4: 20, 5: 17},
 {1: 28, 2: 23, 3: 17, 4: 13, 5: 19})
>>> simulate_game(policy)
({1: 13, 2: 21, 3: 19, 4: 23, 5: 24},
 {1: 22, 2: 20, 3: 19, 4: 19, 5: 20})
>>> simulate_game(policy)
({1: 20, 2: 19, 3: 15, 4: 21, 5: 25},
 {1: 19, 2: 23, 3: 20, 4: 17, 5: 21})
```

If you average over the four example games in listing 10.3, you see that the winners picked 1 an average of 18.75 times per game, whereas the losers picked 1 an average of 22.5 times. This makes sense, because 1 is a bad move. Even though all these games were sampled from the same policy, the distributions differ between the winners and losers, because picking 1 more often caused the agent to lose.

The difference between the moves the agent picks in wins and the moves the agent picks in losses tells you which moves are better. To improve the policy, you can update the probabilities according to those differences. In this case, you can add a small fixed amount for each time a move appears in a win, and subtract a small fixed amount for each time a move appears in a loss. Then the probability distribution will slowly shift toward the moves that appear more often in wins—which you assume are the good moves. For Add It Up, this algorithm works fine. To learn a complicated game like Go, you'll need a more sophisticated scheme for updating the probabilities; we cover that in section 10.2.

Listing 10.4   A policy learning implementation for the simple game Add It Up

```
def normalize(policy):                      Ensures that the policy is a valid
    policy = np.clip(policy, 0, 1)          probability distribution, by making
    return policy / np.sum(policy)          sure it sums to 1

choices = [1, 2, 3, 4, 5]
policy = np.array([0.2, 0.2, 0.2, 0.2, 0.2])      A setting controlling how
learning_rate = 0.0001                            fast you update the policy
for i in range(num_games):
    win_counts, lose_counts = simulate_game(policy)
    for i, choice in enumerate(choices):
        net_wins = win_counts[choice] - lose_counts[choice]
        policy[i] += learning_rate * net_wins
    policy = normalize(policy)            net_wins will be positive if choice
    print('%d: %s' % (i, policy))         appears more often in wins than in losses;
                                          it'll be negative if choice appears more
                                          often in losses than in wins.
```

Figure 10.2 shows how the policy evolves throughout this demo. After about a thousand games, the algorithm learns to stop choosing the worst move. In another thousand games or so, it has more or less arrived at the perfect strategy: selecting 5 on each turn. The curves aren't perfectly smooth. Sometimes the agent will choose a lot of 1s in a game and win anyway; then the policy will (incorrectly) shift toward 1. You rely on these mistakes getting smoothed out over the course of many, many games.

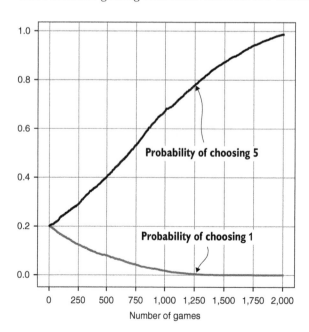

Probability of choosing 5

Probability of choosing 1

Number of games

Figure 10.2   This graph shows how a policy evolves under your simplified policy-learning scheme. Over the course of hundreds of games, the agent gradually becomes less likely to choose the worst move (playing a 1). Likewise, the agent gradually becomes more likely to choose the best move (playing a 5). Both curves are wobbly, because the policy sometimes takes a small step in the wrong direction.

10.2 Modifying neural network policies with gradient descent

There's one glaring difference between learning to play Add It Up and learning to play Go. The policy you used in the Add It Up example doesn't depend on the game state in any way. Picking 5 is always a good move, and picking 1 is always a bad move. In Go, when we say we want to increase the probability of a particular move, we really want to increase the probability of that move *in similar situations*. But the definition of *similar situations* is impossibly vague. Here, we rely on the power of neural networks to tease apart what *similar situations* really means.

When you created a neural network policy in chapter 9, you built a function that took a board position as input and produced a probability distribution over moves as output. For every board position in your experience data, you want to either increase the probability of the chosen move (if it led to a win) or decrease the probability of the chosen move (if it led to a loss). But you can't forcibly modify the probabilities in the policy as you did in section 9.1. Instead, you have to modify the weights of the neural network to make your desired outcome happen. Gradient descent is the tool that makes this possible. Modifying a policy with gradient descent is called *policy gradient learning*. There are a few variations on this idea; the particular learning algorithm we describe in this chapter is sometimes called *Monte Carlo policy gradient*, or the *REINFORCE* method. Figure 10.3 shows the high-level flow of how this process applies to games.

> **NOTE** The *REINFORCE* method stands for *REward Increment = Nonnegative Factor times Offset Reinforcement times Characteristic Eligibility*, which spells out the formula for the gradient update.

Let's recap how supervised learning with gradient descent works, as covered in chapter 5. You chose a loss function that represents how far away your function is from the training data, and calculated its gradient. The goal was to make the value of the loss function smaller, meaning the learned function would match the training data better. Gradient descent—gradually updating the weights in the direction of the gradient of the loss function—provided the mechanism for decreasing the loss function. The gradient told you the direction to shift each weight in order to decrease the loss function.

For policy learning, you want to find the direction to shift each weight in order to bias the policy toward (or away from) a specific move. You can craft a loss function so that its gradient has this property. When you have that, you can take advantage of the fast and flexible infrastructure that Keras provides to modify the weights of your policy network.

Recall the supervised learning you did in chapter 7. For each game state, you also knew the human move that happened in the game. You created a target vector that contained a 0 for each board position, with a 1 to indicate the human move. The loss function measured the gap between the predicted probability distribution, and its gradient indicated the direction to follow to shrink the gap. After completing a batch of gradient descent, the predicted probability of the human move increased slightly.

That's exactly the effect that you want to achieve: increasing the probability of a particular move. For games that your agent won, you can construct the exact same target vector for the agent's move as if it were a human move from a real game record. Then the Keras `fit` function updates the policy in the right direction.

What about the case of a lost game? In that case, you want to decrease the probability of the chosen move, but you don't know the actual best move. Ideally, the update should have the exact opposite effect as in a won game.

As it turns out, if you train with the cross-entropy loss function, you can just insert a −1 into the target vector instead of a 1. This will reverse the sign of the gradient of the loss function, meaning your weights will move in the exact opposite direction, thereby decreasing the probability.

To do this trick, you must use cross-entropy loss; other loss functions, like mean squared error, won't work the same way. In chapter 7, you chose cross-entropy loss because it's the most efficient way to train a network to choose one of a fixed number of options. Here you choose it for a different property: swapping a −1 for a 1 in the target vector will reverse the direction of the gradient.

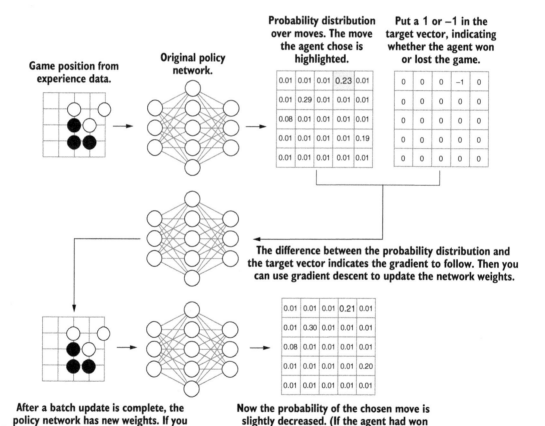

Figure 10.3   A flowchart for policy gradient learning. You start with a collection of game records and their outcomes. For each move the agent chose, you want to either increase the probability of that move (if the agent won the game) or decrease the probability (if the agent lost the game). Gradient descent handles the mechanics of updating the policy weights. After a pass of gradient descent, the probabilities will be shifted in the desired direction.

Recall that your experience data comprises three parallel arrays:

- `states[i]` represents a particular board position your agent faced during self-play.
- `actions[i]` represents the move your agent chose, given that position.
- `rewards[i]` contains a 1 if the agent won the game, and a –1 otherwise.

The following listing implements a `prepare_experience_data` function that packs an experience buffer into a target array suitable for Keras `fit`.

Listing 10.5 Encoding experience data as a target vector

```
def prepare_experience_data(experience, board_width, board_height):
    experience_size = experience.actions.shape[0]
    target_vectors = np.zeros((experience_size, board_width * board_height))
    for i in range(experience_size):
        action = experience.actions[i]
        reward = experience.rewards[i]
        target_vectors[i][action] = reward
    return target_vectors
```

The following listing shows how to implement a `train` function on your `PolicyAgent` class.

Listing 10.6 Training an agent from experience data with policy gradient learning

```
class PolicyAgent(Agent):          lr (learning rate), clipnorm, and batch_size let you fine-tune the
...                                training process; we cover these in detail in the following text.

    def train(self, experience, lr, clipnorm, batch_size):        ◁
        self._model.compile(
            loss='categorical_crossentropy',                      ┐  The compile method
            optimizer=SGD(lr=lr, clipnorm=clipnorm))              │  assigns an optimizer to
                                                                  │  the model; in this case,
        target_vectors = prepare_experience_data(                 │  the stochastic gradient
            experience,                                           ┘  descent (SGD) optimizer.
            self._encoder.board_width,
            self._encoder.board_height)

        self._model.fit(
            experience.states, target_vectors,
            batch_size=batch_size,
            epochs=1)
```

In addition to an experience buffer, this `train` function takes three parameters that modify the behavior of the optimizer:

- `lr` is the *learning rate*, which controls how far to move the weights at each step.
- `clipnorm` provides a hard maximum on how far to move the weights on any individual step.
- `batch_size` controls how many moves from the experience data get combined into a single weight update.

You may need to fine-tune these parameters to get a good result with policy gradient learning. In section 10.3, we provide tips to help you find the right settings.

In chapter 7, you used the Adadelta and Adagrad optimizers, which automatically adapt the learning rate throughout the training process. Unfortunately, they both make assumptions that don't always apply to policy gradient learning. Instead, you should use the basic stochastic gradient descent optimizer and set the learning rate manually. We want to emphasize that 95% of the time, an adaptive optimizer like Adadelta or Adagrad is the best choice; you'll see faster training with fewer headaches. But in rare cases, you need to fall back to plain SGD, and it's good to have some understanding of how to set the learning rate by hand.

Notice also that you run only a single epoch of training on the experience buffer. This is different from chapter 7, where you ran several epochs on the same training set. The key difference is that the training data in chapter 7 was known to be good. Each game move in that data set was a move that a skilled human player chose in a real game. In your self-play data, the game outcomes are partially randomized, and you don't know which moves deserve credit for the wins. You're counting on a huge number of games to smooth out the errors. So you don't want to reuse any single game record: that'll double down on whatever misleading data it contains.

Fortunately, with reinforcement learning, you have an unlimited supply of training data. Instead of running multiple epochs on the same training set, you should run another batch of self-play and generate a new training set.

Now that you have a `train` function ready, the following listing shows a script to do the training. You can find the full script in train_pg.py on GitHub. It consumes experience data files, generated by the self_play script from chapter 9.

> **Listing 10.7   Training on previously saved experience data**

```
learning_agent = agent.load_policy_agent(h5py.File(learning_agent_filename))
for exp_filename in experience_files:                        ◄──────────────┐
    exp_buffer = rl.load_experience(h5py.File(exp_filename))                │
    learning_agent.train(                                                   │
        exp_buffer,              You may have more training data than can fit in
        lr=learning_rate,        memory at once; this implementation will read
        clipnorm=clipnorm,       it in from multiple files, one chunk at a time.
        batch_size=batch_size)
with h5py.File(updated_agent_filename, 'w') as updated_agent_outf:
    learning_agent.serialize(updated_agent_outf)
```

10.3   Tips for training with self-play

Tuning the training process can be difficult. And training a large network is slow, which means you may have to wait a long time to check your results. You should be prepared for some trial and error and a few false starts. This section provides tips for managing a long training process. First, we provide details on how to test and verify your bot's progress. Then we dig into some of the tuning parameters that affect the training process.

Reinforcement learning is slow: if you're training a Go AI, you may need 10,000 self-play games or more before you see a visible improvement. We suggest starting your experiments on a smaller board, such as 9 × 9 or even 5 × 5. On small boards, the game is shorter, so you can generate the self-play games faster; and the game is less complicated, so you need less training data to make progress. That way, you can test your code and tune the training process faster. Once you're confident in your code, you can move up to a larger board size.

10.3.1 Evaluating your progress

In a game as complex as Go, reinforcement learning can take a long time—especially if you don't have access to specialized hardware. Nothing is more frustrating than spending days running a training process, only to find that something went wrong many hours ago. We recommend regularly checking the progress of your learning agent. You do this by simulating more games. The eval_pg_bot.py script pits two versions of your bot against each other; the following listing shows how it works.

> **Listing 10.8   Script for comparing the strength of two agents**

This script tracks wins and losses from the point of view of agent1.

```
wins = 0
losses = 0
color1 = Player.black          ◄─┐  color1 is the color that
for i in range(num_games):          agent1 plays with; agent2
    print('Simulating game %d/%d...' % (i + 1, num_games))
    if color1 == Player.black:   gets the opposite color.
        black_player, white_player = agent1, agent2
    else:
        white_player, black_player = agent1, agent2
    game_record = simulate_game(black_player, white_player)
    if game_record.winner == color1:
        wins += 1                      Swap colors after each game,
    else:                              in case either agent plays
        losses += 1                    better with a particular color.
    color1 = color1.other        ◄─┘
print('Agent 1 record: %d/%d' % (wins, wins + losses))
```

After each batch of training, you can pit the updated agent against the original agent and confirm that the updated agent is improving, or at least not getting worse.

10.3.2 Measuring small differences in strength

After training on thousands of self-play games, your bot may improve by only a few percentage points compared to its predecessor. Measuring a difference that small is fairly difficult. Let's say you've completed a round of training. For evaluation, you run your updated bot against the previous version for 100 games. The updated bot wins 53. Is the new bot truly 3 percentage points stronger? Or did it just get lucky? You need a way to decide whether you have enough data to accurately evaluate your bot's strength.

Imagine that your training did nothing at all, and your updated bot is identical to the previous version. What are the chances that the identical bot wins at least 53 games? Statisticians use a formula called a *binomial test* to calculate that chance. The Python package scipy provides a convenient implementation of the binomial test:

```
>>> from scipy.stats import binom_test
>>> binom_test(53, 100, 0.5)
0.61729941358925255
```

In that snippet:

- 53 represents the number of wins you observed.
- 100 represents the number of games you simulated.
- 0.5 represents the probability of your bot winning a single game if it's identical to its opponent.

The binomial test gave a value of 61.7%. If your bot is truly identical to its opponent, it still has a 61.7% chance of winning 53 or more games. This probability is sometime called a *p-value*. This does *not* mean there's a 61.7% chance that your bot has learned nothing—it just means you don't have enough evidence to judge that. If you want to be confident that your bot has improved, you need to run more trials.

As it turns out, you need quite a few trials to reliably measure a difference in strength this small. If you run 1,000 games and get 530 wins, the binomial test gives a p-value of about 6%. A common guideline is to look for a p-value under 5% before making a decision. But there's nothing magical about that 5% threshold. Instead, think of the p-value as a spectrum indicating how skeptical you should be of your bot's winning record, and use your judgment.

10.3.3 *Tuning a stochastic gradient descent (SGD) optimizer*

The SGD optimizer has a few parameters that can affect its performance. Generally, they have a trade-off between speed and accuracy. Policy gradient learning is typically more sensitive to accuracy than supervised learning, so you'll need to set the parameters appropriately.

The first parameter you must set is the learning rate. To properly set the learning rate, you should understand the problems that an incorrectly set learning rate can cause. Throughout this section, you'll refer to figure 10.4. This shows an imaginary objective function that you're trying to minimize. Conceptually, this diagram is showing the same thing as the diagrams you studied in section 5.4; but here we restrict it to one dimension to illustrate a few specific points. In reality, you're normally optimizing a function over thousands of dimensions.

In chapter 5, you were trying to optimize a loss function that measured the error between your predictions and known correct examples. In this case, the objective is your bot's winning rate. (Technically, in the case of winning percentage, you'd want to maximize the objective instead. The same principles apply in either case, just flipped upside-down.) Unlike the loss function, you can't calculate the winning rate directly, but you can estimate its gradient from the self-play data. In figure 10.4, the x-axis rep-

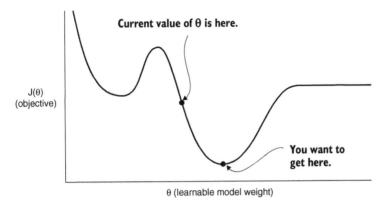

Figure 10.4   This graph shows how a hypothetical objective function may vary with a learnable weight. You want to move the value of θ (Greek letter theta) from its current location to the minimum. You can think of gradient descent as causing the weight to roll downhill.

resents some weight in your network, and the y-axis shows how the value of the objective varies with that weight. The marked point indicates the current state of the network. In the ideal case, you can imagine that gradient descent makes the marked point roll downhill and settle in the valley.

If the learning rate is too small, as in figure 10.5, the optimizer will move the weight in the right direction, but it will take many, many rounds of training to reach the minimum. For the sake of efficiency, you want the learning rate to be as large as possible without causing problems.

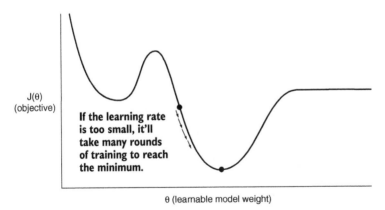

Figure 10.5   In this example, the learning rate is too small, and you need many updates before the weight reaches the minimum.

If you overshoot a little, the objective may not improve as much as it could. But the next gradient will point in the right direction, so it could bounce back and forth for a while, as shown in figure 10.6.

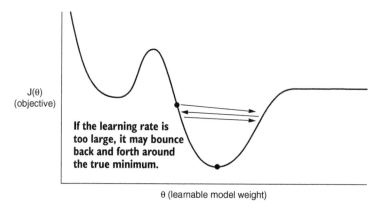

Figure 10.6   Here, the learning rate is too large. The weight overshoots its target. On the next round of learning, the gradient will point in the opposite direction, but it's likely to overshoot again on the next turn. This may cause the weight to bounce back and forth around the true minimum.

In the example objective, if you overshoot a lot, the weight will end up in the flat region on the right. Figure 10.7 shows how that can happen. The gradient is close to zero in that region, which means gradient descent no longer gives you a clue about which direction to move the weight. In that case, the objective can get stuck permanently. This isn't just a theoretical problem: these flat regions are common in networks with rectified linear units, which we introduced in chapter 6. Deep-learning engineers sometimes refer to this problem as *dead ReLUs*: they're considered "dead" because they get stuck returning a value of 0, and stop contributing to the overall learning process.

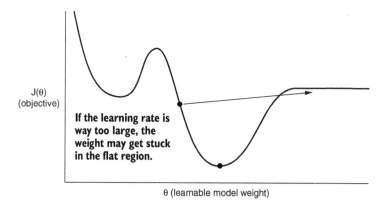

Figure 10.7   In this case, the learning rate is so large that the weight jumped all the way to the flat region on the right. The gradient is 0 in that region, so the optimizer no longer has a clue which direction to go. The weight may be stuck there permanently. This is a common problem in neural networks that use rectified linear units.

Those are the problems that can happen when you overshoot in the correct direction. In policy gradient learning, the problem is even hairier, because you don't know the true gradient you're trying to follow. Somewhere in the universe, there's a theoretical function that relates your agent's playing strength to the weights of its policy network. But you have no way of writing down that function; the best you can do is estimate the gradient from the training data. That estimate is noisy and can sometimes point in the wrong direction. (Recall figure 10.2 from section 10.1; the probability of choosing the best move frequently made a small step in the wrong direction. Self-play data for Go or a similarly complex game will be even noisier than that.)

If you move too far in the wrong direction, the weight may settle in the other valley on the left. Figure 10.8 shows how this case can happen. This is called *forgetting*: the network learned to handle a certain property of the data set, and then suddenly that's undone.

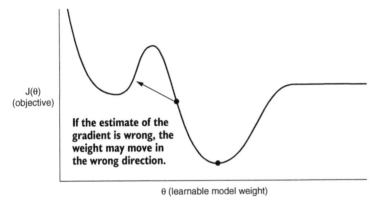

Figure 10.8   In policy gradient learning, you are trying to estimate the true gradient from a very noisy signal. Sometimes a single estimate will point in the wrong direction. If you move the weight too far in the wrong direction, it can jump all the way from the true minimum in the middle to the local minimum on the left, and it may get stuck there a while.

You can also take steps to improve the gradient estimates. Recall that stochastic gradient descent works on mini-batches: the optimizer takes a small subset of the training set, calculates the gradient from just those points, and then updates all the weights. Larger batch sizes tend to smooth out errors. The default batch size in Keras is 32, which is good for many supervised learning problems. For policy learning, we suggest making it a lot larger: try 1,024 or even 2,048 to start.

Finally, policy gradient learning is susceptible to getting stuck in *local maxima*—a case where any incremental change to the policy makes the bot weaker. Sometimes you can escape a local maximum by introducing a little extra randomness to the self-play. Some small fraction of the time (say 1% or 0.5% of turns), the agent can go off-policy and select a completely random move.

Practically speaking, the policy gradient training process goes as follows:

1 Generate a large batch of self-play games (as many as you can fit in memory).
2 Train.
3 Test the updated bot against the previous version.
4 If the bot is measurably stronger, switch to this new version.
5 If the bot is about the same strength, generate more games and train again.
6 If the bot gets significantly weaker, adjust the optimizer settings and retrain.

Tuning the optimizer may feel like threading a needle, but with a little practice and experimentation, you can develop a feel for it. Table 10.1 summarizes the tips we've covered in this section.

Table 10.1 Policy-learning troubleshooting

Symptom	Possible causes	Remedies
Win rate is stuck at 50%.	Learning rate is too small.	Increase learning rate.
	Policy is at a local maximum.	Add more randomness to self-play.
Win rate drops significantly.	Overshoot	Decrease learning rate.
	Bad gradient estimates	Increase batch size.
		Collect more self-play games.

10.4 Summary

- Policy learning is a reinforcement-learning technique that updates a policy from experience data. In the case of game playing, this means updating a bot to choose better moves based on the agent's game results.
- One form of policy learning is to increase the probability of every move that happened in a win, and decrease the probability of every move that happened in a loss. Over thousands of games, this algorithm will slowly update the policy so that it wins more often. This algorithm is *policy gradient learning*.
- *Cross-entropy loss* is a loss function designed for situations where you want to choose one out of a fixed set of options. In chapter 7, you used cross-entropy loss when predicting which move a human would choose in a given game situation. You can also adapt cross-entropy loss for policy gradient learning.
- You can implement policy gradient learning efficiently within the Keras framework by encoding the experience correctly, and then training by using cross-entropy loss.
- Policy gradient training may require hand-tuning the optimizer settings. For policy gradient learning, you may need a smaller learning rate and a larger batch size than you'd use for supervised learning.

Reinforcement learning with value methods

Have you ever read an expert commentary on a high-level chess or Go tournament game? You'll often see comments like, "Black is far behind at this point" or "The result up to here is slightly better for white." What does it mean to be "ahead" or "behind" in the middle of such a strategy game? This isn't basketball, with a running score to refer to. Instead, the commentator means that the board position is favorable to one player or the other. If you want to be precise, you could define it with a thought experiment. Find a hundred evenly matched pairs of players. Give each pair the board position from the middle of the game, and tell them to start playing from there. If the player taking black wins a small majority of the games—say, 55 out of 100—you can say the position was slightly good for black.

Of course, the commentators are doing no such thing. Instead, they're relying on their own intuition, built up over thousands of games, to make a judgment on what might happen. In this chapter, we show how to train a computer game player to make similar judgments. And the computer will learn to do it in much the same way a human will: by playing many, many games.

This chapter introduces the *Q-learning* algorithm. Q-learning is a method for training a reinforcement-learning agent to anticipate how much reward it can expect in the future. (In the context of games, *reward* means *winning games.*) First, we describe how a Q-learning agent makes decisions and improves over time. After that, we show how to implement Q-learning within the Keras framework. Then you'll be ready to train another self-improving game AI, with a different personality from the policy learning agent from chapter 10.

11.1   *Playing games with Q-learning*

Suppose you have a function that tells you your chances of winning after playing a particular move. That's called an *action-value function*—it tells you how valuable a particular action is. Then playing the game would be easy: you'd just pick the highest-value move on each turn. The question, then, is how you come up with an action-value function.

This section describes *Q-learning*, a technique for training an action-value function through reinforcement learning. Of course, you can never learn the true action-value function for moves in Go: that would require reading out the entire game tree, with all its trillions of trillions of trillions of possibilities. But you can iteratively improve an *estimate* of the action-value function through self-play. As the estimate gets more accurate, a bot that relies on the estimate will get stronger.

The name *Q-learning* comes from standard mathematical notation. Traditionally, $Q(s,a)$ is used to represent the action-value function. This is a function of two variables: s represents the state the agent is faced with (for example, a board position); a represents an action the agent is considering (a possible move to play next). Figure 11.1 illustrates the inputs to an action-value function. This chapter focuses on *deep Q-learning*, in which you use a neural network to estimate the Q function. But most of the principles also apply to classical Q-learning, where you approximate the Q function with a simple table that has a row for each possible state and a column for each possible action.

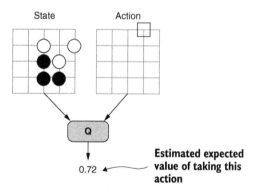

Figure 11.1   An *action-value function* takes two inputs: a state (board position) and an action (proposed move). It returns an estimate of the expected return (chance of winning the game) if the agent chooses this action. The action-value function is traditionally called Q in mathematical notation.

In the previous section, you studied reinforcement learning by directly learning a policy—a rule for selecting moves. The structure of Q-learning will seem familiar. First you make an agent play against itself, recording all the decisions and game results; the game results tell you something about whether the decisions were good; then you update the agent's behavior accordingly. Q-learning differs from policy learning in the way the agent makes decisions in a game, and the way it updates its behavior based on the results.

To build a game-playing agent out of a Q function, you need to turn the Q function into a policy. One option is to plug in every possible move to the Q function and pick the move with the highest expected return, as shown in figure 11.2. This policy is called a *greedy* policy.

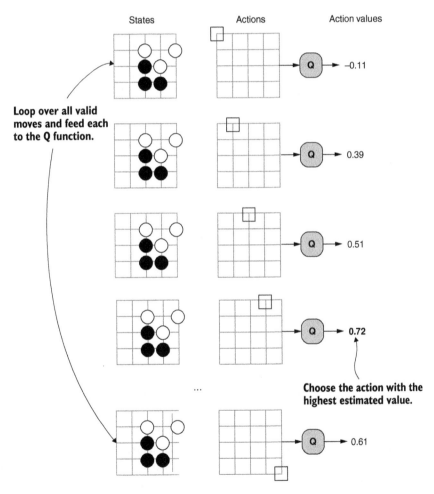

Figure 11.2   In a *greedy* action-value policy, you loop over every possible move and estimate the action value. Then you choose the action with the highest estimated value. (Many legal moves have been omitted to save space.)

If you have confidence in your estimates of the action values, a greedy policy is the best choice. But in order to *improve* your estimates, you need your bot to occasionally explore the unknown. This is called an ε-*greedy* policy: some fraction ε of the time, the policy chooses completely randomly; the rest of the time, it's a normal greedy policy. Figure 11.3 shows this procedure as a flowchart.

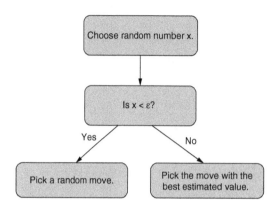

Figure 11.3 Flowchart for the ε-*greedy* action-value policy. This policy tries to balance playing the best move with exploring unknown moves. The value of ε controls that balance.

NOTE ε is the Greek letter epsilon, often used to represent a small fraction.

Listing 11.1 Pseudocode for an ε-greedy policy

```
def select_action(state, epsilon):
    possible_actions = get_possible_actions(state)
    if random.random() < epsilon:                              Random exploration case
        return random.choice(possible_actions)
    best_action = None
    best_value = MIN_VALUE                                     Picking
    for action in get_possible_actions(state):                the best
        action_value = self.estimate_action_value(state, action)   known
        if action_value > best_value:                         move
            best_action = action
            best_value = action_value
    return best_action
```

The choice of ε represents a trade-off. When it's close to 0, the agent chooses whatever the best moves are, according to its current estimate of the action-value; but the agent won't have any opportunity to try new moves and thereby improve its estimates. When ε is higher, the agent will lose more games, but in exchange it'll learn about many unknown moves.

You can make an analogy with the way humans learn a skill, whether it's playing Go or playing the piano. It's common for human learners to hit a plateau—a point where they're comfortable with a certain range of skills but have stopped improving. To get over the hump, you need to force yourself out of your comfort zone and experiment with new things. Maybe that's new fingerings and rhythms on a piano, or new openings and tactics in Go. Your performance may get worse while you find yourself in unfamiliar situations, but after you learn how the new techniques work, you come out stronger than before.

In Q-learning, you generally start with a fairly high value of ε, perhaps 0.5. As the agent improves, you gradually decrease ε. Note that if ε drops all the way to 0, your agent will stop learning: it'll just play out the same game over and over again.

After you've generated a large set of games, the training process for Q-learning is similar to supervised learning. The actions that the agent took provide your training set, and you can treat the game outcomes as known good labels for the data. Of course, there'll be some lucky wins in the training set; but over the course of thousands of games, you can rely on an equal number of losses to cancel them out.

In the same way that the move-prediction model from chapter 7 learned to predict human moves from games it had never seen before, the action-value model can learn to predict the value of moves it has never played before. You can use the game outcomes as the target for the training process, as shown in figure 11.4. To get it to generalize in this way, you need an appropriate neural network design and plenty of training data.

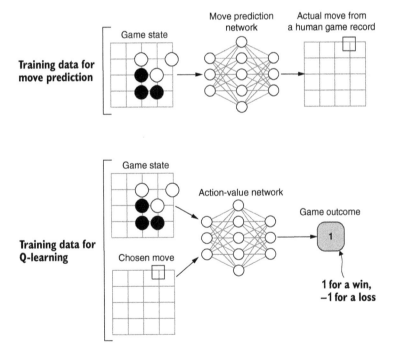

Figure 11.4 Setting up training data for deep Q-learning. At the top, we show how we created training data for the move-prediction networks you used in chapters 6 and 7. The board position was the input, and the actual move was the output. The bottom shows the structure for training data for Q-learning. Both the board position and chosen move are inputs; the output is the game outcome: a 1 for a win and a –1 for a loss.

11.2   *Q-learning with Keras*

This section shows how to implement the Q-learning algorithm in the Keras framework. So far, you've used Keras to train functions that have one input and one output. Because an action-value function has two inputs, you'll need to use new Keras features to design an appropriate network. After introducing two-input networks in Keras, we show how to evaluate moves, assemble training data, and train your agent.

11.2.1   *Building two-input networks in Keras*

In previous chapters, you used the Keras `Sequential` model to define your neural networks. The following listing shows an example model defined with the sequential API.

> **Listing 11.2   Defining a model with the Keras sequential API**

```
from keras.models import Sequential
from keras.layers import Dense

model = Sequential()
model.add(Dense(32, input_shape=(19, 19)))
model.add(Dense(24))
```

Keras provides a second API for defining a neural network: the *functional* API. The functional API provides a superset of the functionality of the sequential API. You can rewrite any sequential network in the functional style, and you can also create complex networks that can't be described in the sequential style.

The main difference is in the way you specify the connections between layers. To connect layers in a sequential model, you repeatedly call `add` on the model object; that automatically connects the last layer's output to the input of the new layer. To connect layers in a functional model, you pass the input layer to the next layer with syntax that looks like a function call. Because you're explicitly creating each connection, you can describe more-complex networks. The following listing shows how to create an identical network to listing 11.2 by using the functional style.

> **Listing 11.3   Defining an identical model with the Keras functional API**

```
from keras.models import Model
from keras.layers import Dense, Input

model_input = Input(shape=(19, 19))
hidden_layer = Dense(32)(model_input)      ◀── Connects model_input to the
output_layer = Dense(24)(hidden_layer)         input of a Dense layer, and
                                               names that layer hidden_layer
model = Model(inputs=[model_input], outputs=[output_layer])
```

Connects hidden_layer to the input of a new Dense layer, and names that layer output_layer

These two models are identical. The sequential API is a convenient way to describe the most common neural networks, and the functional API provides the flexibility to specify multiple inputs and outputs, or complex connections.

Because your action-value network has two inputs and one output, at some point you need to merge the two input chains together. The Keras Concatenate layer lets you accomplish this. A Concatenate layer doesn't do any computation; it just glues together two vectors or tensors into one, as shown in figure 11.5. It takes an optional axis argument that specifies which dimension to concatenate; it defaults to the last dimension, which is what you want in this case. All other dimensions must be the same size.

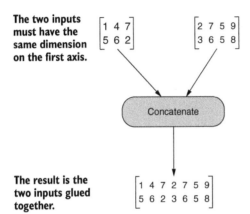

Figure 11.5 The Keras Concatenate layer appends two tensors into one.

Now you can design a network for learning an action-value function. Recall the convolutional networks you used for move prediction in chapters 6 and 7. You can conceptually chunk the network into two steps: first the convolutional layers identify the important shapes of stones on the board; then a dense layer makes a decision based on those shapes. Figure 11.6 shows how the layers in the move-prediction network perform two separate roles.

For the action-value network, you still want to process the board into important shapes and groups of stones. Any shape that's relevant for move prediction is likely to be relevant for estimating the action-value as well, so this part of the network can borrow the same structure. The difference comes in the decision-making step. Instead of making a decision based only on the identified groups of stones, you want to estimate a value based on the processed board *and* the proposed action. So you can bring in the proposed move vector after the convolutional layers. Figure 11.8 illustrates such a network.

Because you use −1 to represent a loss, and 1 to represent a win, the action-value should be a single value in the range of −1 to 1. To achieve this, you add a Dense layer of size 1 with a tanh activation. You may know tanh as the hyperbolic tangent function from trigonometry. In deep learning, you don't care about the trigonometric properties of tanh at all. Instead, you use it because it's a smooth function that's

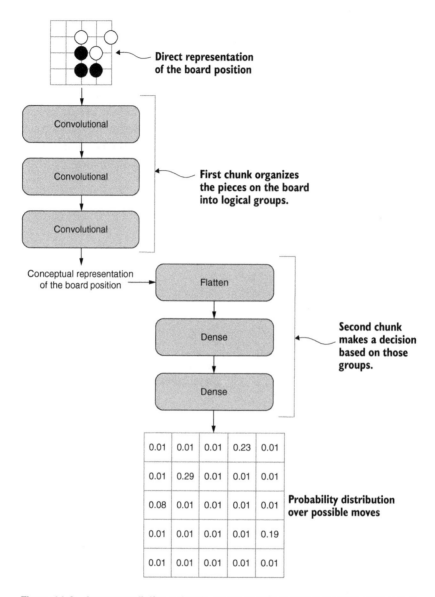

Figure 11.6   A move-prediction network, as covered in chapters 6 and 7. Although there are many layers, you can conceptually think of it as two steps. The convolutional layers process the raw stones and organize them into logical groups and tactical shapes. From that representation, the dense layers can choose an action.

bounded below by –1 and above by 1. No matter what the early layers of the network compute, the output will be in the desired range. Figure 11.7 shows a plot of the tanh function.

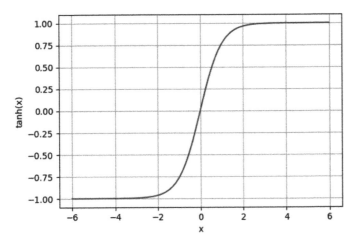

Figure 11.7 The tanh (hyperbolic tangent) function, which clamps its value between –1 and 1

The full specification for your action-value network looks like the following listing.

Listing 11.4 A two-input action-value network

```
from keras.models import Model
from keras.layers import Conv2D, Dense, Flatten, Input
from keras.layers import ZeroPadding2D, concatenate

board_input = Input(shape=encoder.shape(), name='board_input')
action_input = Input(shape=(encoder.num_points(),),
    name='action_input')

conv1a = ZeroPadding2D((2, 2))(board_input)
conv1b = Conv2D(64, (5, 5), activation='relu')(conv1a)

conv2a = ZeroPadding2D((1, 1))(conv1b)
conv2b = Conv2D(64, (3, 3), actionvation='relu')(conv2a)

flat = Flatten()(conv2b)
processed_board = Dense(512)(flat)

board_and_action = concatenate([action_input, processed_board])
hidden_layer = Dense(256, activation='relu')(board_and_action)
value_output = Dense(1, activation='tanh')(hidden_layer)

model = Model(inputs=[board_input, action_input],
    outputs=value_output)
```

Add as many convolutional layers as you like. Anything that worked well for move prediction ought to work well here.

You may want to experiment with the size of this hidden layer.

The tanh activation layer clamps the output between –1 and 1.

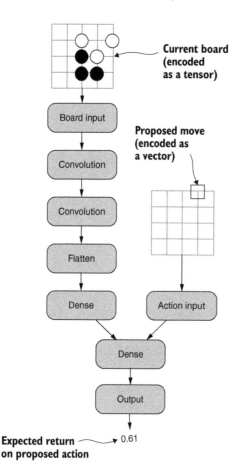

Current board (encoded as a tensor)

Proposed move (encoded as a vector)

Expected return on proposed action → 0.61

Figure 11.8   The two-input neural network described in listing 11.4. The game board goes through several convolutional layers, just like the move-prediction network from chapter 7. The proposed move goes into a separate input. The proposed move is combined with the output of the convolutional layers and passed through another dense layer.

11.2.2 Implementing the ε-greedy policy with Keras

Let's start building a QAgent that can learn via Q-learning; this code will live in the dlgo/rl/q.py module. Listing 11.5 shows the constructor: just as in our policy learning agent, it takes a model and a board encoder. You also define two utility methods. The set_temperature method lets you change the value of ε, which you'll want to vary across the training process. Just as in chapter 9, the set_collector method lets you attach an ExperienceCollector object to store the experience data for later training.

Listing 11.5   Constructor and utility methods for a Q-learning agent

```
class QAgent(Agent):
    def __init__(self, model, encoder):
        self.model = model
        self.encoder = encoder
        self.collector = None
        self.temperature = 0.0
```

```
def set_temperature(self, temperature):
    self.temperature = temperature
```
temperature is the ε value that controls
how randomized the policy is.

```
def set_collector(self, collector):
    self.collector = collector
```
See chapter 9 for more information about using a
collector object to record the agent's experiences.

Next, you implement the ε-greedy policy. Instead of just picking the top-rated move,
you sort all the moves and try them in order. As in chapter 9, this prevents the agent
from self-destructing at the end of a won game.

Listing 11.6 Selecting moves for a Q-learning agent

```
class QAgent(Agent):
    ...
    def select_move(self, game_state):
        board_tensor = self.encoder.encode(game_state)

        moves = []
        board_tensors = []
        for move in game_state.legal_moves():
            if not move.is_play:
                continue
            moves.append(self.encoder.encode_point(move.point))
            board_tensors.append(board_tensor)
        if not moves:
            return goboard.Move.pass_turn()

        num_moves = len(moves)
        board_tensors = np.array(board_tensors)
        move_vectors = np.zeros(
            (num_moves, self.encoder.num_points()))
        for i, move in enumerate(moves):
            move_vectors[i][move] = 1

        values = self.model.predict(
            [board_tensors, move_vectors])
        values = values.reshape(len(moves))

        ranked_moves = self.rank_moves_eps_greedy(values)

        for move_idx in ranked_moves:
            point = self.encoder.decode_point_index(
                moves[move_idx])
            if not is_point_an_eye(game_state.board,
                                   point,
                                   game_state.next_player):
                if self.collector is not None:
                    self.collector.record_decision(
                        state=board_tensor,
                        action=moves[move_idx],
                    )
                return goboard.Move.play(point)
        return goboard.Move.pass_turn()
```

Generates a
list of all
valid moves

If there are no valid moves
left, the agent can just pass.

Values will
be an *N* × 1
matrix,
where *N* is
the number
of legal
moves; the
reshape call
converts to
a vector of
size *N*.

One-hot encodes all the valid
moves (see chapter 5 for
more on one-hot encoding).

This is the two-input form of predict:
you pass the two inputs as a list.

Ranks
the moves
according to
the ε-greedy
policy

Picks the first non-
self-destructive move
in your list, similar to
the self-play agents
from chapter 9

You'll fall through here
if all the valid moves
are determined to be
self-destructive.

Records the decision in an
experience buffer; see chapter 9

Q-learning and tree search

The structure of the `select_move` implementation is similar to some of the tree-search algorithms we covered in chapter 4. For example, alpha-beta search relies on a board evaluation function: a function that takes a board position and estimates which player is ahead and by how much. This is similar, but not identical, to the action-value function we've covered in this chapter. Suppose the agent is playing as black and evaluating some move X. It gets an estimated action-value of 0.65 for X. Now, you know exactly what the board will look like after playing move X, and you know that a win for black is a loss for white. So you can say that the next board position has a value of –0.65 for white.

Mathematically, we describe the relationship as follows:

$$Q(s,a) = -V(s')$$

where s' is the state that white sees after black chooses move a.

Although Q-learning in general can apply to any environment, this equivalence between the action-value of one state and value of the next state is true only in deterministic games.

Chapter 12 covers a third reinforcement-learning technique that includes learning a value function directly, instead of an action-value function. Chapters 13 and 14 show methods for integrating such a value function with a tree-search algorithm.

All that remains is the code to sort the moves in order from most valuable to least valuable. The complication is that you have two parallel arrays: `values` and `moves`. The NumPy `argsort` function provides a convenient way to handle this. Instead of sorting an array in place, `argsort` returns a list of indices. Then you can read off the elements of the parallel array according to those indices. Figure 11.9 illustrates how `argsort` works. Listing 11.7 shows how to rank the moves by using `argsort`.

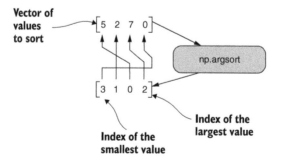

Figure 11.9 An illustration of the `argsort` function from the NumPy library. `argsort` takes a vector of values that you want to sort. Instead of sorting the values directly, it returns a vector of indices that will give you the values in sorted order. So the first value of the output vector is the index of the smallest value in the input, and the last value of the output vector is the index of the largest value in the input.

Listing 11.7 Selecting moves for a Q-learning agent

```
class QAgent(Agent):
    ...
    def rank_moves_eps_greedy(self, values):
        if np.random.random() < self.temperature:
            values = np.random.random(values.shape)
        ranked_moves = np.argsort(values)
        return ranked_moves[::-1]
```

In the exploration case, rank the moves by random numbers instead of the real values.

Gets the indices of the moves in order from least value to highest value

The [::-1] syntax is the most efficient way to reverse a vector in NumPy. This returns the moves in order from highest value to least value.

With this in place, you're ready to generate self-play games with your Q-learning agent. Next, we cover how to train the action-value network.

11.2.3 Training an action-value function

After you have a batch of experience data, you're ready to update the agent's network. With policy gradient learning, you knew the approximate gradient that you wanted, but you had to come up with a complicated scheme to apply that gradient update inside the Keras framework. In contrast, training with Q-learning is a straightforward application of the Keras `fit` function. You can directly put the game results into the target vector.

Chapter 6 covered two loss functions: mean squared error and cross-entropy loss. You used cross-entropy loss when you wanted to match one out of a discrete set of items: in that case, you were trying to match one of the points on a Go board. The Q function, on the other hand, has a continuous value that can be anywhere in the range –1 to 1. For this problem, we prefer mean squared error.

The following listing shows the implementation of a `train` function for the QAgent class.

Listing 11.8 Training the Q-learning agent from its experience

lr and batch_size are options to fine-tune the training process. See chapter 10 for more discussion.

```
class QAgent(Agent):
    ...
    def train(self, experience, lr=0.1, batch_size=128):
        opt = SGD(lr=lr)
        self.model.compile(loss='mse', optimizer=opt)

        n = experience.states.shape[0]
        num_moves = self.encoder.num_points()
        y = np.zeros((n,))
        actions = np.zeros((n, num_moves))
        for i in range(n):
            action = experience.actions[i]
            reward = experience.rewards[i]
            actions[i][action] = 1
            y[i] = reward
```

mse is mean squared error. You use mse instead of categorical_crossentropy because you're trying to learn a continuous value.

```
self.model.fit(
    [experience.states, actions], y,   ⟵── Passes the two different inputs as a list
    batch_size=batch_size,
    epochs=1)
```

11.3 Summary

- An *action-value function* estimates how much reward an agent can expect after taking a specific action. In the case of games, this means the expected chance of winning.

- *Q-learning* is the technique of reinforcement learning by estimating an action-value function (traditionally notated as Q).

- While training a Q-learning agent, you normally use an *ε-greedy policy*. Under this policy, the agent will select the highest-valued move a fraction of the time, and a random move the rest of the time. The parameter ε controls how much the agent will explore unknown moves in order to learn more about them.

- The Keras functional API lets you design neural networks with multiple inputs, multiple outputs, or complex internal connections. For Q-learning, you can use the functional API to build a network with separate inputs for the game state and the proposed move.

Reinforcement learning with actor-critic methods

This chapter covers

- Using advantage to make reinforcement learning more efficient
- Making a self-improving game AI with the actor-critic method
- Designing and training multi-output neural networks in Keras

If you're learning to play Go, one of the best ways to improve is to get a stronger player to review your games. Sometimes the most useful feedback just points out where you won or lost the game. The reviewer might give comments like, "You were already far behind by move 30" or "At move 110, you had a winning position, but your opponent turned it around by move 130."

Why is this feedback helpful? You may not have time to scrutinize all 300 moves in a game, but you can focus your full attention on a 10- or 20-move sequence. The reviewer lets you know which parts of the game are important.

Reinforcement-learning researchers apply this principle in *actor-critic learning*, which is a combination of policy learning (as covered in chapter 10) and value learning (as covered in chapter 11). The policy function plays the role of the *actor*: it picks what moves to play. The value function is the *critic*: it tracks whether the agent is ahead or behind in the course of the game. That feedback guides the training process, in the same way that a game review can guide your own study.

This chapter describes how to make a self-improving game AI with actor-critic learning. The key concept that makes it all work is *advantage*, the difference between the actual game outcome and the expected outcome. We start by illustrating how advantage can improve the training process. After that, we're ready to build an actor-critic game agent. First we show how to implement move selection; then we implement the new training process. In both functions, we borrow heavily from the code examples in chapters 10 and 11. The end result is the best of both worlds: it combines the benefits of policy learning and Q-learning into one agent.

12.1   *Advantage tells you which decisions are important*

In chapter 10, we briefly mentioned the credit-assignment problem. Suppose your learning agent played a game with 200 moves and ultimately won the game. Because it won, you can assume it chose at least a few good moves, but it probably chose a couple of bad moves as well. *Credit assignment* is the problem of separating the good moves, which you want to reinforce, from the bad moves, which you should ignore. This section introduces the concept of *advantage*, a formula for estimating how much a particular decision contributed to the final result. First we describe how advantage helps with credit assignment; then we provide code samples showing how to calculate it.

12.1.1   *What is advantage?*

Imagine you're watching a basketball game; while the fourth quarter ticks down, your favorite player nails a three-pointer. How excited do you get? It depends on the game state. If the score is 80 to 78, you're probably jumping out of your seat. If the score is 110 to 80, you're indifferent. What's the difference? In a close game, a three-point swing creates a huge change in the expected outcome of the game. On the other hand, if the game is a blowout, a single play won't affect the result. The most important plays happen while the outcome is still in doubt. In reinforcement learning, advantage is a formula that quantifies this concept.

To calculate advantage, you first need an estimate of the value of a state, which we denote as $V(s)$. This is the expected return the agent will see, given that it has already arrived at a particular state s. In games, you can think of $V(s)$ as indicating whether the board position is good for black or white. If $V(s)$ is close to 1, your agent is in a favorable position; if $V(s)$ is close to -1, your agent is losing.

If you recall the action-value function $Q(s,a)$ from the previous chapter, the concept is similar. The difference is that $V(s)$ represents how favorable the board is *before* you choose a move; $Q(s,a)$ represents how favorable the board is *after* you choose a move.

The definition of advantage is usually specified as follows:

$$A = Q(s,a) - V(s)$$

One way to think of this is that if you're in a good state (that is, $V(s)$ is high), but you make a terrible move ($Q(s,a)$ is low), you give away your advantage: hence the calculation is negative. One problem with this formula, however, is that you don't know how to calculate $Q(s,a)$. But you can consider the reward you get at the end of the game as an unbiased estimate of the true Q. So you can wait until you get your reward R, and then estimate the advantage as follows:

$$A = R - V(s)$$

That's the calculation you'll use to estimate advantage throughout this chapter. Let's see how this value is useful.

For the purposes of illustration, you'll pretend that you already have an accurate way to estimate $V(s)$. In reality, your agent learns its value-estimating function and its policy function simultaneously. The next section covers how that works.

Let's work through a few examples:

- At the beginning of a game, $V(s) = 0$: both players have a roughly equal chance. Suppose your agent wins the game; then its reward will be 1, so the advantage of its first move is $1 - 0 = 1$.
- Imagine that the game is almost over and your agent has practically locked the game up, so $V(s) = 0.95$. If your agent does indeed win the game, the advantage from that state is $1 - 0.95 = 0.05$.
- Now imagine your agent has another winning position, where once again $V(s) = 0.95$. But in this game, your bot somehow blunders away the end game and loses, giving it a reward of -1. Its advantage from that state is $-1 - 0.95 = -1.95$.

Figures 12.1 and 12.2 illustrate the advantage calculation for a hypothetical game. In this game, your learning agent slowly pulled ahead over the first few moves; then it made some big mistakes and fell all the way to a lost position. Somewhere before move 150, it suddenly managed to reverse the game and finally cruised to a win. Under the policy gradient technique from chapter 10, you'd weight each move equally in this game. With actor-critic learning, you want to find the most important moves and give them greater weight. The advantage calculation shows you how.

Because the learning agent won, the advantage is given by $A(s) = 1 - V(s)$. In figure 12.2, you can see that the advantage curve has the same shape as the estimated value curve, but flipped upside down. The largest advantage comes while the agent was far behind. Because most players would lose in such a bad situation, the agent must have made a great move somewhere.

After the agent had already pulled back ahead, around move 160 or so, its decisions are no longer interesting: the game had already wrapped up. The advantage in that section is close to 0.

Later in this chapter, we show how to adjust the training process based on the advantages. Before that, you need to calculate and store advantage through your self-play process.

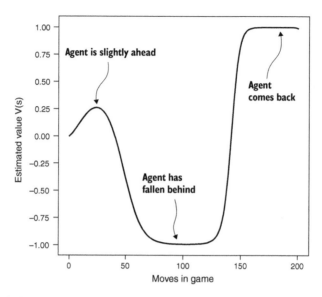

Figure 12.1   Estimated values over the course of a hypothetical game. This game lasted 200 moves. In the beginning, the learning agent pulled slightly ahead; then it fell far behind; then it suddenly reversed the game and came out with a win.

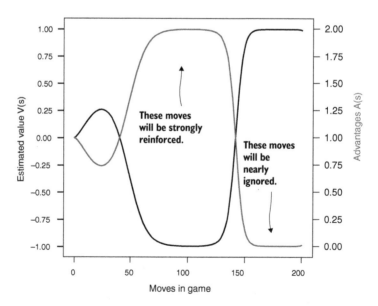

Figure 12.2   The advantages for each move in a hypothetical game. The learning agent won the game, so its final reward was 1. The moves that led to the comeback have an advantage close to 2, so they'll be strongly reinforced during training. The moves near the end of the game, when the outcome was already decided, have an advantage close to 0, so they'll be nearly ignored during training.

12.1.2 *Calculating advantage during self-play*

To calculate advantage, you'll update your `ExperienceCollector` that you defined in chapter 9. Originally, an experience buffer tracked three parallel arrays: states, actions, and rewards. You can add a fourth parallel array to track advantages. To fill this array, you need both the estimated value for each state and the final game outcome. You won't have the latter until the end; so in the middle of the episode, you can accumulate estimated values, and when the game is complete, you can translate those into advantages.

Listing 12.1 Updating `ExperienceCollector` to track advantages

```
class ExperienceCollector:
    def __init__(self):
        self.states = []
        self.actions = []                          These can span
        self.rewards = []                          many episodes.
        self.advantages = []
        self._current_episode_states = []
        self._current_episode_actions = []         These are reset at the
        self._current_episode_estimated_values = []  end of every episode.
```

Similarly, you need to update the `record_decision` method to accept an estimated value along with a state and an action.

Listing 12.2 Updating `ExperienceCollector` to store estimated values

```
class ExperienceCollector:
...
    def record_decision(self, state, action,
            estimated_value=0):
        self._current_episode_states.append(state)
        self._current_episode_actions.append(action)
        self._current_episode_estimated_values.append(
            estimated_value)
```

Then, in the `complete_episode` method, you can calculate the advantage of each decision the agent made.

Listing 12.3 Calculating advantage at the end of an episode

```
class ExperienceCollector:
...
    def complete_episode(self, reward):
        num_states = len(self._current_episode_states)
        self.states += self._current_episode_states
        self.actions += self._current_episode_actions
        self.rewards += [reward for _ in range(num_states)]

        for i in range(num_states):
            advantage = reward - \                      Calculates the advantage
                self._current_episode_estimated_values[i]  of each decision
            self.advantages.append(advantage)
```

```
        self._current_episode_states = []
        self._current_episode_actions = []
        self._current_episode_estimated_values = []
```
Reset the per-episode buffers.

You also need to update the `ExperienceBuffer` class and `combine_experience` helper to handle the advantages.

Listing 12.4 Adding advantage to the `ExperienceBuffer` structure

```
class ExperienceBuffer:
    def __init__(self, states, actions, rewards, advantages):
        self.states = states
        self.actions = actions
        self.rewards = rewards
        self.advantages = advantages

    def serialize(self, h5file):
        h5file.create_group('experience')
        h5file['experience'].create_dataset('states',
data=self.states)
        h5file['experience'].create_dataset('actions',
data=self.actions)
        h5file['experience'].create_dataset('rewards',
data=self.rewards)
        h5file['experience'].create_dataset('advantages',
data=self.advantages)


def combine_experience(collectors):
    combined_states = np.concatenate(
[np.array(c.states) for c in collectors])
    combined_actions = np.concatenate(
[np.array(c.actions) for c in collectors])
    combined_rewards = np.concatenate(
[np.array(c.rewards) for c in collectors])
    combined_advantages = np.concatenate([
        np.array(c.advantages) for c in collectors])

    return ExperienceBuffer(
        combined_states,
        combined_actions,
        combined_rewards,
        combined_advantages)
```

Your experience classes are now ready to track advantage. You can still use these classes with techniques that don't rely on advantage; just ignore the contents of the advantages buffer while training.

12.2 *Designing a neural network for actor-critic learning*

Chapter 11 showed how to define a neural network with two inputs in Keras. The Q-learning network had one input for the board and one input for the proposed move. For actor-critic learning, you want a network with one input and two outputs. The input is a representation of the board state. One output is a probability distribution over moves—the actor. The other output represents the expected return from the current position—the critic.

Building a network with two outputs brings a surprising bonus: each output serves as a sort of regularizer on the other. (Recall from chapter 6 that *regularization* is any technique to prevent your model from *overfitting* to the exact data set it was trained on.) Imagine that a group of stones on the board is in danger of getting captured. This fact is relevant for the value output, because the player with the weak stones is probably behind. It's also relevant to the action output, because you probably want to either attack or defend the weak stones. If your network learns a "weak stone" detector in the early layers, that's relevant to both outputs. Training on both outputs forces the network to learn a representation that's useful for both goals. This can often improve generalization and sometimes even speed up training.

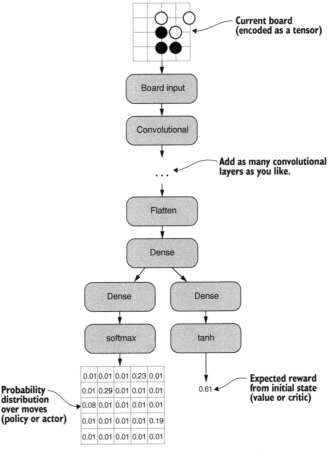

Figure 12.3 A neural network suitable for actor-critic learning for Go. This network has a single input, which takes a representation of the current board position. The network produces two outputs. One output indicates which moves it should play—this is the policy output, or the actor. The other output indicates which player is ahead in the game—this is the value output, or the critic. The critic isn't used in playing a game but helps the training process.

Chapter 11 introduced the Keras functional API, which gives you full freedom to connect layers in your network however you like. You'll use it again here to build the network described in figure 12.3; this code goes in the init_ac_agent.py script.

Listing 12.5   A two-output network with a policy output and a value output

```python
from keras.models import Model
from keras.layers import Conv2D, Dense, Flatten, Input

board_input = Input(shape=encoder.shape(), name='board_input')

conv1 = Conv2D(64, (3, 3),
               padding='same',
               activation='relu')(board_input)
conv2 = Conv2D(64, (3, 3),
               padding='same',
               activation='relu')(conv1)
conv3 = Conv2D(64, (3, 3),
               padding='same',
               activation='relu')(conv2)

flat = Flatten()(conv3)
processed_board = Dense(512)(flat)

policy_hidden_layer = Dense(
    512, activation='relu')(processed_board)
policy_output = Dense(
    encoder.num_points(), activation='softmax')(
    policy_hidden_layer)

value_hidden_layer = Dense(
    512, activation='relu')(
    processed_board)
value_output = Dense(1, activation='tanh')(
    value_hidden_layer)

model = Model(inputs=board_input,
  outputs=[policy_output, value_output])
```

> **Add as many convolutional layers as you like.**

> **This example uses hidden layers of size 512. Experiment to find the best size. The three hidden layers don't need to be the same size.**

> **This output yields the policy function.**

> **This output yields the value function.**

This network has three convolutional layers with 64 filters each. That's on the smaller side for a Go-playing network, but it has the advantage of faster training. As always, we encourage you to experiment with different network structures here.

The policy output represents a probability distribution over possible moves. The dimension is equal to the number of points on the board, and you use the softmax activation to ensure that the policy sums to 1.

The value output is a single number in the range of –1 to 1. This output has dimension 1, and you use a tanh activation to clamp the value.

12.3   *Playing games with an actor-critic agent*

Selecting moves is almost exactly the same as in the policy agent from chapter 10. You make two changes. First, because the model now produces two outputs, you need a little extra code to unpack the results. Second, you need to pass the estimated value to the experience collector, along with the state and action. The process of picking a move from the probability distribution is identical. The following listing shows the updated select_move implementation. We've called out places where it differs from implementation of chapter 10's policy agent.

Listing 12.6 Selecting a move for an actor-critic agent

```
class ACAgent(Agent):
...
    def select_move(self, game_state):
        num_moves = self.encoder.board_width * \
self.encoder.board_height

        board_tensor = self.encoder.encode(game_state)
        X = np.array([board_tensor])

        actions, values = self.model.predict(X)
        move_probs = actions[0]
        estimated_value = values[0][0]

        eps = 1e-6
        move_probs = np.clip(move_probs, eps, 1 - eps)
        move_probs = move_probs / np.sum(move_probs)

        candidates = np.arange(num_moves)
        ranked_moves = np.random.choice(
            candidates, num_moves, replace=False, p=move_probs)
        for point_idx in ranked_moves:
            point = self.encoder.decode_point_index(point_idx)
            move = goboard.Move.play(point)
            move_is_valid = game_state.is_valid_move(move)
            fills_own_eye = is_point_an_eye(
                game_state.board, point,
game_state.next_player)
            if move_is_valid and (not fills_own_eye):
                if self.collector is not None:
                    self.collector.record_decision(
                        state=board_tensor,
                        action=point_idx,
                        estimated_value=estimated_value
                    )
                return goboard.Move.play(point)
        return goboard.Move.pass_turn()
```

> **Because this is a two-output model, predict returns a tuple containing two NumPy arrays.**

> **predict is a batch call that can process several boards at once, so you must select the first element of the array to get the probability distribution you want.**

> **The values are represented as a one-dimensional vector, so you must pull out the first element to get the value as a plain float.**

> **Include the estimated value in the experience buffer.**

12.4 Training an actor-critic agent from experience data

Training your actor-critic network looks like a combination of training the policy network in chapter 10 and the action-value network in chapter 11. To train a two-output network, you construct a separate training target for each output, and choose a separate loss function for each output. This section describes how to convert the experience data to training targets, and how to use the Keras fit function with multiple outputs.

Recall how you encoded training data for policy gradient learning. For any game position, the training target was a vector the same size as the board, with a 1 or –1 in the slot corresponding to the chosen move; the 1 indicated a win, and the –1 indicated a loss. In your actor-critic learning, you use the same encoding scheme for the training data, but you replace the 1 or –1 with the advantage of the move. The advantage will have the same sign as the final reward, so the probability of the game decision will move in the same direction as in simple policy learning. But it'll move further for actions that were deemed important, and move just a little for actions with an advantage that's close to zero.

For the value output, the training target is the total reward. This looks exactly like the training target for Q-learning. Figure 12-4 illustrates the training setup.

When you have multiple outputs in a network, you can pick a different loss function for each output. You'll use categorical cross-entropy for the policy output, and mean squared error for the value output. (Refer to chapters 10 and 11 for an explanation of why those loss functions make sense for those purposes.)

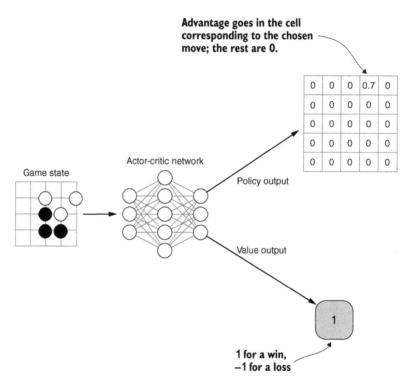

Figure 12.4   Training setup for actor-critic learning. The neural network has two outputs: one for the policy and one for the value. Each gets its own training target. The policy output is trained against a vector the same size as the board. The cell in the vector corresponding to the chosen move is filled in with the advantage calculated for that move; the rest are zero. The value output is trained against the final outcome of the game.

One new Keras feature you'll use is *loss weights*. By default, Keras will sum the loss function for each output to get the overall loss function. If you specify loss weights, Keras will scale each individual loss function before summing. This allows you to adjust the relative importance of each output. In our experiments, we found the value loss was large compared to the policy loss, so we scaled down the value loss by half. Depending on your exact network and training data, you may need to adjust the loss weights somewhat.

> **TIP** Keras will print out the computed loss values every time you call `fit`. For a two-output network, it'll print out the two losses separately. You can check there to see whether the magnitudes are comparable. If one loss is far larger than the other, consider adjusting the weights. Don't worry about getting too precise.

The following listing shows how to encode the experience data as training data, and then call `fit` on the training targets. The structure is similar to the `train` implementations from chapters 10 and 11.

Listing 12.7 Selecting a move for an actor-critic agent

lr (learning rate) and batch_size are tuning parameters for the optimizer; refer to chapter 10 for more discussion.

categorical_crossentropy is for the policy output, just as in chapter 10. mse (mean squared error) is for the value output, just as in chapter 11. The order here matches the order in the Model constructor in listing 12.5.

The weight 1.0 applies to the policy output; the weight 0.5 applies to the value output.

This is the same as the encoding scheme used in chapter 11.

This is the same as the encoding scheme used in chapter 10, but weighted by the advantage.

```
class ACAgent(Agent):
...
    def train(self, experience, lr=0.1, batch_size=128):
        opt = SGD(lr=lr)
        self.model.compile(
            optimizer=opt,
            loss=['categorical_crossentropy', 'mse'],
            loss_weights=[1.0, 0.5])

        n = experience.states.shape[0]
        num_moves = self.encoder.num_points()
        policy_target = np.zeros((n, num_moves))
        value_target = np.zeros((n,))
        for i in range(n):
            action = experience.actions[i]
            policy_target[i][action] = experience.advantages[i]
            reward = experience.rewards[i]
            value_target[i] = reward

        self.model.fit(
            experience.states,
            [policy_target, value_target],
            batch_size=batch_size,
            epochs=1)
```

Now that you have all the pieces, let's try actor-critic learning end-to-end. You'll start with a 9 × 9 bot so you can see results faster. The cycle will go like this:

1 Generate self-play games in chunks of 5,000.

2 After each chunk, train the agent and compare it to the previous version of your bot.

3 If the new bot can beat the previous bot 60 out of 100 games, you've successfully improved your agent! Start the process over with the new bot.

4 If the updated bot wins fewer than 60 out of 100 games, generate another chunk of self-play games and retrain. Continue training until the new bot is strong enough.

The benchmark of 60 wins out of 100 is somewhat arbitrary; it's a nice round number that gives you reasonable confidence that your bot is truly stronger, and not just lucky.

Start by initializing a bot with the init_ac_agent script (as shown in listing 12.5):

```
python init_ac_agent.py --board-size 9 ac_v1.hdf5
```

After this, you should have a new file, ac_v1.hdf5, that contains the weights for your new bot. At this point, both the bot's play and its value estimates are essentially random. You can now start generating self-play games:

```
python self_play_ac.py \
--board-size 9 \
--learning-agent ac_v1.hdf5 \
--num-games 5000 \
--experience-out exp_0001.hdf5
```

If you're not fortunate enough to have access to a fast GPU, this is a good time to go out for a coffee or take the dog for a walk. When the self_play script is done, the output will look something like this:

```
Simulating game 1/5000...
 9 ooxxxxxxx
 8 ooox.xx.x
 7 oxxxxooxx
 6 oxxxxxox.
 5 oooooxoxx
 4 ooo.oooxo
 3 ooooooooo
 2 .oo.ooo.o
 1 ooooooooo
   ABCDEFGHJ
W+28.5
...
Simulating game 5000/5000...
 9 x.x.xxxxx
 8 xxxxx.xxx
 7 .x.xxxxoo
 6 xxxx.xo.o
 5 xxxxxooo
 4 xooooooxo
```

```
 3 xoooxxxxo
 2 o.o.oxxxx
 1 ooooox.x.
   ABCDEFGHJ
B+15.5
```

After this, you should have an exp_0001.hdf5 file containing a big chunk of game records. The next step is to train:

```
python train_ac.py \
--learning-agent bots/ac_v1.hdf5 \
--agent-out bots/ac_v2.hdf5 \
--lr 0.01 --bs 1024 \
exp_0001.hdf5
```

This will take the neural network currently stored in ac_v1.hdf1, run a single epoch of training against the data in exp_0001.hdf, and save the updated agent to ac_v2.hdf5. The optimizer will use a learning rate of 0.01 and a batch size of 1,024. You should see output something like this:

```
Epoch 1/1
574234/574234 [==============================] - 15s 26us/step - loss:
➡ 1.0277 - dense_3_loss: 0.6403 - dense_5_loss: 0.7750
```

Notice that the loss is now broken into two values: dense_3_loss and dense_5_loss, corresponding to the policy output and the value output, respectively.

After this, you can compare the updated bot against its predecessor with the eval_ac_bot.py script:

```
python eval_ac_bot.py \
--agent1 bots/ac_v2.hdf5 \
--agent2 bots/ac_v1.hdf5 \
--num-games 100
```

The output should look something like this:

```
...
Simulating game 100/100...
 9 oooxxxxx.
 8 .oox.xxxx
 7 ooxxxxxxx
 6 .oxx.xxxx
 5 oooxxx.xx
 4 o.ox.xx.x
 3 ooxxxxxxx
 2 ooxx.xxxx
 1 oxxxxxxx.
   ABCDEFGHJ
B+31.5
Agent 1 record: 60/100
```

In this case, the output shows you exactly hit the threshold of 60 wins out of 100: you can have reasonable confidence that your bot has learned something useful. (This is just example output, of course; your actual results will look a little different, and that's fine.) Because the ac_v2 bot is measurably stronger than ac_v1, you can switch to generating games with ac_v2:

```
python self_play_ac.py \
--board-size 9 \
--learning-agent ac_v2.hdf5 \
--num-games 5000 \
--experience-out exp_0002.hdf5
```

When that's done, you can train and evaluate again:

```
python train_ac.py \
--learning-agent bots/ac_v2.hdf5 \
--agent-out bots/ac_v3.hdf5 \
--lr 0.01 --bs 1024 \
exp_0002.hdf5
python eval_ac_bot.py \
--agent1 bots/ac_v3.hdf5 \
--agent2 bots/ac_v2.hdf5 \
--num-games 100
```

This case wasn't quite as successful as the last time:

```
Agent 1 record: 51/100
```

The ac_v3 bot beat the ac_v2 bot only 51 times out of 100. With those results, it's hard to say whether ac_v3 is a tiny bit stronger or not; the safest conclusion is that it's basically the same strength as ac_v2. But don't despair. You can generate more training data and try again:

```
python self_play_ac.py \
--board-size 9 \
--learning-agent ac_v2.hdf5 \
--num-games 5000 \
--experience-out exp_0002a.hdf5
```

The train_ac script will accept multiple training data files on the command line:

```
python train_ac.py \
--learning-agent ac_v2.hdf5 \
--agent-out ac_v3.hdf5 \
--lr 0.01 --bs 1024 \
exp_0002.hdf5 exp_0002a.hdf5
```

After each additional batch of games, you can compare against ac_v2 again. In our experiments, we needed three batches of 5,000 games—a total of 15,000 games—before we got a satisfactory result:

```
Agent 1 record: 62/100
```

Success! With 62 wins against ac_v2, you can feel confident that ac_v3 is stronger than ac_v2. Now you can switch over to generating self-play games with ac_v3, and repeat the cycle again.

It's unclear exactly how strong a Go bot can get with this actor-critic implementation alone. We've shown that you can train a bot to learn basic tactics, but its strength is bound to top out at some point. By deeply integrating reinforcement learning with a kind of tree search, you *can* train a bot that's stronger than any human player; chapter 14 covers that technique.

12.5 Summary

- *Actor-critic learning* is a reinforcement-learning technique in which you simultaneously learn a policy function and a value function. The policy function tells you how to make decisions, and the value function helps improve the training process for the value function. You can apply actor-critic learning to the same kinds of problems where you'd apply policy gradient learning, but actor-critic learning is often more stable.

- *Advantage* is the difference between the actual reward an agent sees and the expected reward at some point in the episode. For games, this is the difference between the actual game result (win or loss) and the expected value (as estimated by the agent's value model).

- Advantage helps identify the important decisions in a game. If a learning agent wins a game, the advantage will be largest for moves it made from an even or losing position. The advantage will be close to zero for moves it made after the game was already decided.

- A Keras sequential network can have multiple outputs. In actor-critic learning, this lets you create a single network to model both the policy function and the value function.

Part 3

Greater than the sum of its parts

At this point, you've learned a number of AI techniques that draw from classical tree search, machine learning, and reinforcement learning. Each is powerful on its own, but each also has limitations. To make a truly strong Go AI, you'll need to combine everything you've learned so far. Integrating all these pieces is a serious engineering feat. This part covers the architecture of AlphaGo, the AI that rocked the Go world—and the AI world! To conclude the book, you'll learn about the elegantly simple design of AlphaGo Zero, the strongest version of AlphaGo to date.

AlphaGo: Bringing it all together

This chapter covers

- Diving into the guiding principles that led Go bots to play at superhuman strength
- Using tree search, supervised deep learning, and reinforcement learning to build such a bot
- Implementing your own version of DeepMind's AlphaGo engine

When DeepMind's Go bot AlphaGo played move 37 of game 2 against Lee Sedol in 2016, it took the Go world by storm. Commentator Michael Redmond, a professional player with nearly a thousand top-level games under his belt, did a double-take on air; he even briefly removed the stone from the demo board while looking around as if to confirm that AlphaGo made the right move. ("I still don't really understand the mechanics of it," Redmond told the American Go E-Journal the next day.) Lee, the world-wide dominant player of the past decade, spent 12 minutes studying the board before responding. Figure 13.1 displays the legendary move.

The move defied conventional Go theory. The diagonal approach, or *shoulder hit*, is an invitation for the white stone to extend along the side and make a solid wall. If

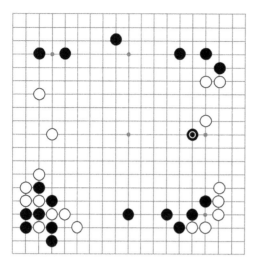

Figure 13.1   The legendary shoulder hit that AlphaGo played against Lee Sedol in the second game of their series. This move stunned many professional players.

the white stone is on the third line and the black stone is on the fourth line, this is considered a roughly even exchange: white gets points on the side, while black gets influence toward the center. But when the white stone is on the fourth line, the wall locks up too much territory. (To any strong Go players who are reading, we apologize for drastically oversimplifying this.) A fifth-line shoulder hit looks a little amateurish—or at least it did until "Professor Alpha" took four out of five games against a legend. The shoulder hit was the first of many surprises from AlphaGo. Fast-forward a year, and everyone from top pros to casual club players is experimenting with AlphaGo moves.

In this chapter, you're going to learn how AlphaGo works by implementing all of its building blocks. AlphaGo is a clever combination of supervised deep learning from professional Go records (which you learned about in chapters 5–8), deep reinforcement learning with self-play (covered in chapters 9–12), and using these deep networks to improve tree search in a novel way. You might be surprised by how much you already know about the ingredients of AlphaGo. To be more precise, the AlphaGo system we'll be describing in detail works as follows:

- You start off by training *two* deep convolutional neural networks (*policy networks*) for move prediction. One of these network architectures is a bit deeper and *produces more-accurate results*, whereas the other one is smaller and *faster to evaluate*. We'll call them the *strong* and *fast* policy networks, respectively.

- The strong and fast policy networks use a slightly more sophisticated board encoder with 48 feature planes. They also use a deeper architecture than what you've seen in chapters 6 and 7, but other than that, they should look familiar. Section 13.1 covers AlphaGo's policy network architectures.

- After the first training step of policy networks is complete, you take the strong policy network as a starting point for self-play in section 13.2. If you do this with a lot of compute power, this will result in a massive improvement of your bots.

- As a next step, you take the strong self-play network to derive a *value network* from it in section 13.3. This completes the network training stage, and you don't do any deep learning after this point.

- To play a game of Go, you use tree search as a basis for play, but instead of plain Monte Carlo rollouts as in chapter 4, you use the fast policy network to guide the next steps. Also, you balance the output of this tree-search algorithm with what your value function tells you. We'll tell you all about this innovation in section 13.4.

- Performing this whole process from training policies, to self-play, to running games with search on a superhuman level requires massive compute resources and time. Section 13.5 gives you some ideas on what it took to make AlphaGo as strong as it is and what to expect from your own experiments.

Figure 13.2 gives an overview of the whole process we just sketched. Throughout the chapter, we'll zoom into parts of this diagram and provide you with more details in the respective sections.

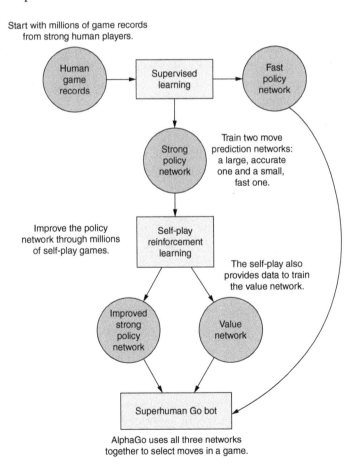

Figure 13.2 How to train the three neural networks that power the AlphaGo AI. Starting with a collection of human game records, you can train two neural networks to predict the next move: a small, fast network and a large, strong network. You can then further improve the playing strength of the large network through reinforcement learning. The self-play games also provide data to train a value network. AlphaGo then uses all three networks in a tree-search algorithm that can produce incredibly strong game play.

13.1   *Training deep neural networks for AlphaGo*

In the introduction, you learned that AlphaGo uses three neural networks: two policy networks and one value network. Although this may seem like a lot at first, in this section, you'll see that these networks and the input features that feed into them are conceptually close to each other. Perhaps the most surprising part about deep learning as used in AlphaGo is how much you already know about it after completing chapters 5 to 12. Before we go into details of how these neural networks are built and trained, let's quickly discuss their role in the AlphaGo system:

- *Fast policy network*—This Go move-prediction network is comparable in size to the networks you trained in chapters 7 and 8. Its purpose isn't to be the most accurate move predictor, but rather a good predictor that's really fast at predicting moves. This network is used in section 13.4 in tree-search rollouts—and you've seen in chapter 4 that you need to create a lot of them quickly for tree search to become an option. We'll put a little less emphasis on this network and focus on the following two.

- *Strong policy network*—This move-prediction network is optimized for accuracy, not speed. It's a convolutional network that's deeper than its fast version and can be more than twice as good at predicting Go moves. As the fast version, this network is trained on human game-play data, as you did in chapter 7. After this training step is completed, the strong policy network is used as a starting point for self-play by using reinforcement-learning techniques from chapters 9 and 10. This step will make this policy network even stronger.

- *Value network*—The self-play games played by the strong policy network generate a new data set that you can use to train a value network. Specifically, you use the outcome of these games and the techniques from chapters 11 and 12 to learn a value function. This value network will then play an integral role in section 13.4.

13.1.1   *Network architectures in AlphaGo*

Now that you roughly know what each of the three deep neural networks is used for in AlphaGo, we can show you how to build these networks in Python, using Keras. Here's a quick description of the network architectures, before we show you the code. If you need a refresher on terminology for convolutional networks, have a look at chapter 7 again.

- The strong policy network is a 13-layer convolutional network. All of these layers produce 19 × 19 filters; you consistently keep the original board size across the whole network. For this to work, you need to *pad* the inputs accordingly, as you did in chapter 7. The first convolutional layer has a kernel size of 5, and all following layers work with a kernel size of 3. The last layer uses softmax activations and has one output filter, and the first 12 layers use ReLU activations and have 192 output filters each.

- The value network is a 16-layer convolutional network, the first 12 of which are *exactly the same as the strong policy network.* Layer 13 is an additional convolutional layer, structurally identical to layers 2–12. Layer 14 is a convolutional layer with kernel size 1 and one output filter. The network is topped off with two dense layers, one with 256 outputs and ReLU activations, and a final one with one output and tanh activation.

As you can see, both policy and value networks in AlphaGo are the same kind of deep convolutional neural network that you already encountered in chapter 6. The fact that these two networks are so similar allows you to define them in a single Python function. Before doing so, we introduce a little shortcut in Keras that shortens the network definition quite a bit. Recall from chapter 7 that you can pad input images in Keras with the ZeroPadding2D utility layer. It's perfectly fine to do so, but you can save some ink in your model definition by moving the padding into the Conv2D layer. What you want to do in both value and policy networks is to pad the input to each convolutional layer so that the output filters have the *same* size as the input (19 × 19). For instance, instead of explicitly padding the 19 × 19 input of the first layer to 23 × 23 images so that the following convolutional layer with kernel size 5 produces 19 × 19 output filters, you tell the convolutional layer to retain the input size. You do this by providing the argument padding='same' to your convolutional layer, which will take care of the padding for you. With this neat shortcut in mind, let's define the first 11 layers that AlphaGo's policy and value networks have in common. You find this definition in our GitHub repository in alphago.py in the dlgo.networks module.

Listing 13.1 Initializing a neural network for both policy and value networks in AlphaGo

```
from keras.models import Sequential
from keras.layers.core import Dense, Flatten
from keras.layers.convolutional import Conv2D
```

> **With this Boolean flag, you specify whether you want a policy or value network.**

```
def alphago_model(input_shape, is_policy_net=False,
                  num_filters=192,
                  first_kernel_size=5,
                  other_kernel_size=3):
```

All but the last convolutional layers have the same number of filters.

> **The first layer has kernel size 5, all others only 3.**

```
    model = Sequential()
    model.add(
        Conv2D(num_filters, first_kernel_size, input_shape=input_shape,
        padding='same',
               data_format='channels_first', activation='relu'))

    for i in range(2, 12):
        model.add(
            Conv2D(num_filters, other_kernel_size, padding='same',
                   data_format='channels_first', activation='relu'))
```

The first 12 layers of AlphaGo's policy and value network are identical.

Note that you didn't yet specify the input shape of the first layer. That's because that shape differs slightly for policy and value networks. You'll see the difference when we introduce the AlphaGo board encoder in the next section. To continue the definition of model, you're just one final convolutional layer away from defining the strong policy network.

Listing 13.2   Creating AlphaGo's strong policy network in Keras

```
if is_policy_net:
    model.add(
        Conv2D(filters=1, kernel_size=1, padding='same',
               data_format='channels_first', activation='softmax'))
    model.add(Flatten())
    return model
```

As you can see, you add a final `Flatten` layer to flatten the predictions and ensure consistency with your previous model definitions from chapters 5 to 8.

If you want to return AlphaGo's value network instead, adding two more `Conv2D` layers, two `Dense` layers, and one `Flatten` layer to connect them will do the job.

Listing 13.3   Building AlphaGo's value network in Keras

```
else:
    model.add(
        Conv2D(num_filters, other_kernel_size, padding='same',
               data_format='channels_first', activation='relu'))
    model.add(
        Conv2D(filters=1, kernel_size=1, padding='same',
               data_format='channels_first', activation='relu'))
    model.add(Flatten())
    model.add(Dense(256, activation='relu'))
    model.add(Dense(1, activation='tanh'))
    return model
```

We don't explicitly discuss the architecture of the fast policy network here; the definition of input features and network architecture of the fast policy is technically involved and doesn't contribute to a deeper understanding of the AlphaGo system. For your own experiments, it's perfectly fine to use one of the networks from our dlgo.networks module, such as `small`, `medium`, or `large`. The main idea for the fast policy is to have a smaller network than the strong policy that's quick to evaluate. We'll guide you through the training process in more detail throughout the next sections.

13.1.2 *The AlphaGo board encoder*

Now that you know all about the network architectures used in AlphaGo, let's discuss how to encode Go board data the AlphaGo way. You've implemented quite a few board encoders in chapters 6 and 7 already, including `oneplane`, `sevenplane`, or `simple`, all of which you stored in the dlgo.encoders module. The feature planes used in

AlphaGo are just a little more sophisticated than what you've encountered before, but represent a natural continuation of the encoders shown so far.

The AlphaGo board encoder for policy networks has 48 feature planes; for value networks, you augment these features with one additional plane. These 48 planes are made up of 11 concepts, some of which you've used before and others that are new. We'll discuss each of them in more detail. In general, AlphaGo makes a bit more use of Go-specific tactical situations than the board encoder examples we've discussed so far. A prime example of this is making the concept of *ladder captures and escapes* (see figure 13.3) part of the feature set.

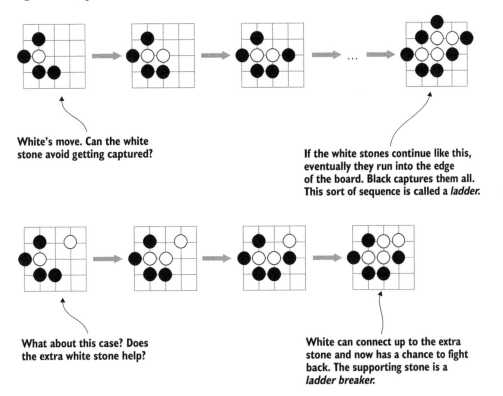

White's move. Can the white stone avoid getting captured?

If the white stones continue like this, eventually they run into the edge of the board. Black captures them all. This sort of sequence is called a *ladder.*

What about this case? Does the extra white stone help?

White can connect up to the extra stone and now has a chance to fight back. The supporting stone is a *ladder breaker.*

Figure 13.3   AlphaGo encoded many Go tactical concepts directly into its feature planes, including *ladders.* **In the first example, a white stone has just one liberty—meaning black could capture on the next turn. The white player extends the white stone to gain an extra liberty. But black can again reduce the white stones to one liberty. This sequence continues until it hits the edge of the board, where white is captured. On the other hand, if there's a white stone in the path of the ladder, white may be able to escape capture. AlphaGo included feature planes that indicated whether a ladder would be successful.**

A technique you consistently used in all of your Go board encoders that's also present in AlphaGo is the use of *binary features.* For instance, when capturing liberties (empty adjacent points on the board), you didn't just use one feature plane with liberty counts for each stone on the board, but chose a binary representation with planes indicating whether a stone had 1, 2, 3, or more liberties. In AlphaGo, you see the exact same idea, but with eight feature planes to binarize counts. In the example of liberties, that means eight planes to indicate 1, 2, 3, 4, 5, 6, 7, or at least 8 liberties for a stone.

The only fundamental difference from what you've seen in chapters 6 to 8 is that AlphaGo encodes stone color explicitly in *separate* feature planes. Recall that in the sevenplane encoder from chapter 7, you had liberty planes for both black and white stones. In AlphaGo, you have only one set of features counting liberties. Additionally, all features are expressed in terms of the player to play next. For instance, the feature set Capture Size, counting the number of stones that would be captured by a move, counts the stones the *current* player would capture, whatever stone color this might be.

Table 13.1 summarizes all the features used in AlphaGo. The first 48 planes are used for policy networks, and the last one only for value networks.

Table 13.1   Feature planes used in AlphaGo

Feature name	Number of planes	Description
Stone color	3	Three feature planes indicating stone color—one each for the current player, the opponent, and the empty points on the board.
Ones	1	A feature plane entirely filled with the value 1.
Zeros	1	A feature plane entirely filled with the value 0.
Sensibleness	1	A move on this plane is 1 if the move is legal and doesn't fill the current player's eyes, and 0 otherwise.
Turns since	8	This set of eight binary planes indicates how many moves ago a move was played.
Liberties	8	Number of liberties of the string of stones this move belongs to, split into eight binary planes.
Liberties after move	8	If this move was played, how many liberties would this result in?
Capture size	8	How many opponent stones would this move capture?
Self-atari size	8	If this move was played, how many of your own stones would be put into atari and could be captured by the opponent in the next move?
Ladder capture	1	Can this stone be captured in a ladder?
Ladder escape	1	Can this stone escape all possible ladders?
Current player color	1	A plane filled with 1s if current player is black, or 0s if the player is white.

The implementation of these features can be found in our GitHub repository under alphago.py in the dlgo.encoders module. Although implementing each of the feature sets from table 13.1 isn't difficult, it's also not particularly interesting when compared to all the exciting parts making up AlphaGo that still lie ahead of us. Implementing ladder captures is somewhat tricky, and encoding the number of turns since a move was played requires modifications to your Go board definition. So if you're interested in how this can be done, check out our implementation on GitHub.

Let's quickly look at how an `AlphaGoEncoder` can be initialized, so you can use it to train deep neural networks. You provide a Go board size and a Boolean called `use_player_plane` that indicates whether to use the 49th feature plane. This is shown in the following listing.

Listing 13.4 Signature and initialization of your AlphaGo board encoder

```
class AlphaGoEncoder(Encoder):
    def __init__(self, board_size, use_player_plane=False):
        self.board_width, self.board_height = board_size
        self.use_player_plane = use_player_plane
        self.num_planes = 48 + use_player_plane
```

13.1.3 *Training AlphaGo-style policy networks*

Having network architectures and input features ready, the first step of training policy networks for AlphaGo follows the exact procedure we introduced in chapter 7: specifying a board encoder and an agent, loading Go data, and training the agents with this data. Figure 13.4 illustrates the process. The fact that you use slightly more elaborate features and networks doesn't change this one bit.

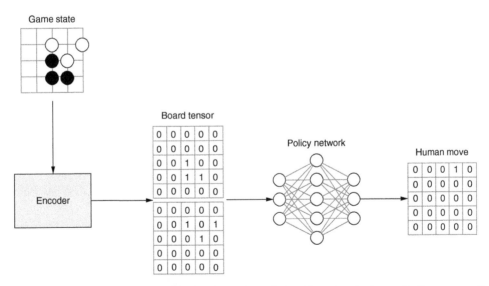

Figure 13.4   The supervised training process for AlphaGo's policy networks is exactly the same as the flow covered in chapters 6 and 7. You replay human game records and reproduce the game states. Each game state is encoded as a tensor (this diagram shows a tensor with only two planes; AlphaGo used 48 planes). The training target is a vector the same size as the board, with a 1 where the human actually played.

To initialize and train AlphaGo's strong policy network, you first need to instantiate an AlphaGoEncoder, and create two Go data generators for training and testing, just as you did in chapter 7. You find this step on GitHub under examples/alphago/ alphago _policy_sl.py.

Listing 13.5   Loading data for the first step of training AlphaGo's policy network

```
from dlgo.data.parallel_processor import GoDataProcessor
from dlgo.encoders.alphago import AlphaGoEncoder
from dlgo.agent.predict import DeepLearningAgent
from dlgo.networks.alphago import alphago_model

from keras.callbacks import ModelCheckpoint
import h5py

rows, cols = 19, 19
num_classes = rows * cols
num_games = 10000

encoder = AlphaGoEncoder()
processor = GoDataProcessor(encoder=encoder.name())
generator = processor.load_go_data('train', num_games, use_generator=True)
test_generator = processor.load_go_data('test', num_games,
➥ use_generator=True)
```

Next, you can load AlphaGo's policy network by using the alphago_model function defined earlier in this section and compile this Keras model with categorical cross-entropy and stochastic gradient descent. We call this model alphago_sl_policy to signify that it's a policy network trained by supervised learning (*sl*).

Listing 13.6   Creating an AlphaGo policy network with Keras

```
input_shape = (encoder.num_planes, rows, cols)
alphago_sl_policy = alphago_model(input_shape, is_policy_net=True)

alphago_sl_policy.compile('sgd', 'categorical_crossentropy',
➥ metrics=['accuracy'])
```

Now all that's left for this first stage of training is to call fit_generator on this policy network, using both training and test generators as you did in chapter 7. Apart from using a larger network and a more sophisticated encoder, this is precisely what you did in chapters 6 to 8.

After training has finished, you can create a DeepLearningAgent from model and encoder and store it for the next two training phases that we discuss next.

Listing 13.7   Training and persisting a policy network

```
epochs = 200
batch_size = 128
alphago_sl_policy.fit_generator(
```

```
        generator=generator.generate(batch_size, num_classes),
        epochs=epochs,
        steps_per_epoch=generator.get_num_samples() / batch_size,
        validation_data=test_generator.generate(batch_size, num_classes),
        validation_steps=test_generator.get_num_samples() / batch_size,
        callbacks=[ModelCheckpoint('alphago_sl_policy_{epoch}.h5')]
)

alphago_sl_agent = DeepLearningAgent(alphago_sl_policy, encoder)

with h5py.File('alphago_sl_policy.h5', 'w') as sl_agent_out:
    alphago_sl_agent.serialize(sl_agent_out)
```

For the sake of simplicity, in this chapter you don't need to train fast and strong policy networks separately, as in the original AlphaGo paper. Instead of training a smaller and faster second policy network, you can use alphago_sl_agent as the fast policy. In the next section, you'll see how to use this agent as a starting point for reinforcement learning, which will lead to a stronger policy network.

13.2 Bootstrapping self-play from policy networks

Having trained a relatively strong policy agent with alphago_sl_agent, you can now use this agent to let it play against itself, using the policy gradient algorithm covered in chapter 10. As you'll see in section 13.5, in DeepMind's AlphaGo you let *different versions of the strong policy network* play against the currently strongest version. This prevents overfitting and results in overall better performance, but our simple approach of letting alphago_sl_agent play against itself conveys the general idea to use self-play to make a policy agent stronger.

For the next training phase, you first load the supervised-learning policy network alphago_sl_agent twice: one version serves as your new reinforcement-learning agent called alphago_rl_agent, and the other as its opponent. This step can be found under examples/alphago/alphago_policy_sl.py on GitHub.

> **Listing 13.8 Loading the trained policy network twice to create two self-play opponents**

```
from dlgo.agent.pg import PolicyAgent
from dlgo.agent.predict import load_prediction_agent
from dlgo.encoders.alphago import AlphaGoEncoder
from dlgo.rl.simulate import experience_simulation
import h5py

encoder = AlphaGoEncoder()

sl_agent = load_prediction_agent(h5py.File('alphago_sl_policy.h5'))
sl_opponent = load_prediction_agent(h5py.File('alphago_sl_policy.h5'))

alphago_rl_agent = PolicyAgent(sl_agent.model, encoder)
opponent = PolicyAgent(sl_opponent.model, encoder)
```

Next, you can use these two agents to engage in self-play and store the resulting experience data, for training purposes. This experience data is used to train alphago_rl_agent with it. You then store the trained reinforcement-learning policy agent and the experience data acquired through self-play, because you need this data to train AlphaGo's value network with it as well.

Listing 13.9 Generating self-play data for your `PolicyAgent` to learn from

```
num_games = 1000
experience = experience_simulation(num_games, alphago_rl_agent, opponent)

alphago_rl_agent.train(experience)

with h5py.File('alphago_rl_policy.h5', 'w') as rl_agent_out:
    alphago_rl_agent.serialize(rl_agent_out)

with h5py.File('alphago_rl_experience.h5', 'w') as exp_out:
    experience.serialize(exp_out)
```

Note that this example uses a utility function called `experience_simulation` from dlgo.rl.simulate. The implementation can be found on GitHub, but all this function does is set up two agents to engage in self-play for a specified number of games (num_games) and return the experience data as `ExperienceCollector`, a concept introduced in chapter 9.

When AlphaGo entered the stage in 2016, the strongest open source Go bot was *Pachi* (which you can learn more about in appendix C), ranked around 2 dan amateur level. Simply letting the reinforcement learning agent alphago_rl_agent pick the next move led to an impressive 85% win rate of AlphaGo against Pachi. Convolutional neural networks were used for Go move prediction before, but never fared better than around 10% against Pachi. This shows you the relative strength gain of self-play over purely supervised learning with deep neural networks. If you run your own experiments, don't expect your bots to start out on such a high ranking—it's unlikely that you have (or can afford) the compute power necessary.

13.3 *Deriving a value network from self-play data*

The third and last step in AlphaGo's network-training process is to train a value network *from the same self-play experience data* that you just used for alphago_rl_agent. This step looks structurally similar to the last. You first initialize an AlphaGo value network and create a `ValueAgent` with an AlphaGo board encoder. This training step can also be found in examples/alphago/alphago_value.py on GitHub.

Listing 13.10 Initializing an AlphaGo value network

```
from dlgo.networks.alphago import alphago_model
from dlgo.encoders.alphago import AlphaGoEncoder
from dlgo.rl import ValueAgent, load_experience
import h5py

rows, cols = 19, 19
encoder = AlphaGoEncoder()
```

```
input_shape = (encoder.num_planes, rows, cols)
alphago_value_network = alphago_model(input_shape)

alphago_value = ValueAgent(alphago_value_network, encoder)
```

You can now pick up the experience data from self-play once again and train your value agent with it, after which you persist the agent just as the other two before.

Listing 13.11   Training a value network from experience data

```
experience = load_experience(h5py.File('alphago_rl_experience.h5', 'r'))

alphago_value.train(experience)

with h5py.File('alphago_value.h5', 'w') as value_agent_out:
    alphago_value.serialize(value_agent_out)
```

At this point, if you were to break into the premises of DeepMind's AlphaGo team (you shouldn't) and assume team members used Keras in the same way you did to train AlphaGo (they didn't), then getting your hands on the network parameters for fast policy, strong policy, and value network, you'd have a Go bot that plays at super-human level. That is, provided you know how to use these three deep networks appropriately in a tree search algorithm. The next section is all about that.

13.4   *Better search with policy and value networks*

Recall from chapter 4 that in pure Monte Carlo tree search applied to the game of Go, you build a tree of game states by using these four steps:

1   *Select*—You traverse the game tree by randomly selecting among *children.*
2   *Expand*—You add a new *node* to the tree (a new game state).
3   *Evaluate*—From this state, which is sometimes referred to as a *leaf*, simulate a game completely randomly.
4   *Update*—After the simulation is completed, update your tree statistics accordingly.

Simulating many games will lead to more and more accurate statistics, which you can then use to pick the next move.

The AlphaGo system uses a more sophisticated tree-search algorithm, but you'll still recognize many of its parts. The preceding four steps are still integral to AlphaGo's MCTS algorithm, but you'll use deep neural networks in a smart way to evaluate positions, expand nodes, and track statistics. For the rest of the chapter, we'll show you exactly how and develop a version of AlphaGo's tree search along the way.

13.4.1   *Using neural networks to improve Monte Carlo rollouts*

Sections 13.1, 13.2, and 13.3 described in detail how to train three neural networks for AlphaGo: fast and strong policies and a value network. How can you use these networks to improve Monte Carlo tree search? The first thing that comes to mind is to stop playing games at random and instead use a policy network to guide rollouts. That's exactly what the fast policy network is for and it explains the name—rollouts need to be *fast* to carry out many of them.

The following listing shows how to greedily select moves from a policy network for a given Go game state. You choose the best possible move until the game is over and return 1 if the current player wins, and –1 otherwise.

> **Listing 13.12   Doing rollouts with the fast policy network**

```
def policy_rollout(game_state, fast_policy):
    next_player = game_state.next_player()
    while not game_state.is_over():
        move_probabilities = fast_policy.predict(game_state)
        greedy_move = max(move_probabilities)
        game_state = game_state.apply_move(greedy_move)

    winner = game_state.winner()
    return 1 if winner == next_player else -1
```

Using this rollout strategy is already beneficial in itself, because policy networks are naturally much better at choosing moves than tossing a coin. But you still have a lot of room for improvement.

For instance, when you arrive at a leaf node in your tree and need to expand it, instead of randomly selecting a new node for expansion, you can *ask the strong policy network for good moves*. A policy network gives you a probability distribution over all next moves, and each node can track this probability so that strong moves (according to the policy) are chosen more likely than others. We call these node probabilities *prior probabilities*, because they give us prior knowledge of the strength of a move before doing any tree search.

Finally, here's how the value network comes into play. You already improved your rollout mechanism by replacing random guesses with a policy network. Nevertheless, at each leaf you compute only the outcome of a single game to estimate how valuable the leaf is. But estimating the value of a position is precisely what you trained the value network to be good at, so you already have a sophisticated guess for that. What AlphaGo does is *weigh* the outcome of rollouts against the output of the value network. If you think about it, that's similar to the way you make decisions when playing games as a human: you try to look ahead as many moves as realistically possible, but you also take your experience of the game into account. If you can read out a sequence of moves that might be good for you, that can supersede your hunch that the position isn't that great, and vice versa.

Now that you roughly know what each of the three deep neural networks used in AlphaGo are for and how tree search can be improved with them, let's have a closer look at the details.

13.4.2 *Tree search with a combined value function*

In chapter 11, you saw action values, also called *Q-values*, applied to the game of Go. To recap, for a current board state s and a potential next move a, an action value $Q(s,a)$ estimates how good a move a would be in the situation s. You'll see in a bit how to define $Q(s,a)$; for now, just note that each node in an AlphaGo search tree stores Q-values. Additionally, each node tracks *visit counts*, meaning how often this node has been traversed by search, as well as *prior probabilities* $P(s,a)$, or how valuable the strong policy network thinks action a would be from s.

Each node in a tree has precisely one parent, but potentially multiple *children*, which you can encode as Python dictionary-mapping moves to other nodes. With this convention, you can define an `AlphaGoNode` as follows.

Listing 13.13   A simple view on a node in an AlphaGo tree

```
class AlphaGoNode:
    def __init__(self, parent, probability):
        self.parent = parent
        self.children = {}

        self.visit_count = 0
        self.q_value = 0
        self.prior_value = probability
```

Let's say you're being thrown into an ongoing game, have already built a large tree, and collected visit counts and a good estimate of action values. What you want is to simulate a number of games and track game statistics so that at the end of the simulation, you can pick the best move you found. How do you traverse the tree to simulate a game? If you're in game state s and denote the respective visit count as $N(s)$, you can choose an action as follows:

$$a' = \operatorname{argmax}_a Q(s,a) + \frac{P(s,a)}{1 + N(s,a)}$$

This might look a little complicated at first, but you can break down this equation:

- The argmax notation means that you take the argument a for which the formula $Q(s,a) + P(s,a) / (1 + N(s,a))$ is maximized.
- The term you maximize is composed of two parts, the Q-value and prior probabilities *normalized* by visit counts.
- In the beginning, visit counts are zeros, which means you give equal weight to Q-value and prior probability by maximizing over $Q(s,a) + P(s,a)$.
- If visit counts become very large, the term $P(s,a) / (1 + N(s,a))$ becomes negligible, which effectively leaves you with $Q(s,a)$.
- We call this utility function $u(s,a) = P(s,a) / (1 + N(s,a))$. In the next section, you'll slightly modify $u(s,a)$, but this version has all the components you need to reason about it. With this notation, you can also write a' = $\operatorname{argmax}_a Q(s,a) + u(s,a)$ for move selection.

To summarize, you select actions by weighing prior probabilities against Q-values. As you traverse the tree, accumulate visit counts, and get better estimates of Q, you slowly forget about your *prior estimation* and put more and more trust in Q-values. You could also say that you rely less on prior knowledge and explore more. This might be analogous to your own game-play experience. Let's say you play your favorite strategy board game all night. At the beginning of the night, you bring all your prior experience to the table, but as the night progresses, you (hopefully) try new things and update your beliefs about what works and what doesn't.

So this is how AlphaGo *selects* moves from an existing tree, but how about *expanding* the tree if you've reached a leaf *l*? See figure 13.5. First, you compute the predictions of the strong policy network $P(l)$ and store them as prior probabilities for each child of *l*. Then you *evaluate* the leaf node by *combining policy rollouts and value network* as follows:

$$V(l) = \lambda \cdot \text{value}(l) + (1 - \lambda) \cdot \text{rollout}(l)$$

In this equation, value(l) is the result of your value network for *l*, rollout(l) denotes the game result of a fast policy rollout from *l*, and λ is a value between 0 and 1, which you will set to 0.5 by default.

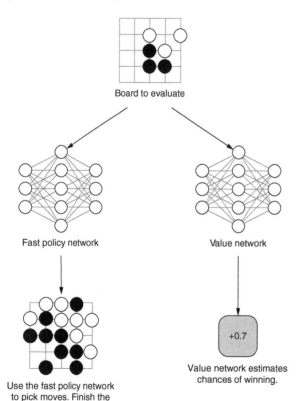

Board to evaluate

Fast policy network   Value network

Use the fast policy network
to pick moves. Finish the
game and see who wins.

+0.7

Value network estimates
chances of winning.

Figure 13.5   To evaluate possible board positions, AlphaGo combines two independent evaluations. First, it feeds the board position to its value network, which directly returns an estimated chance of winning. Second, it uses the fast policy network to complete the game from that position and observes who won. The evaluation used in the tree search is a weighted sum of those two parts.

Taking a step back, keep in mind that you want to simulate a total of *n* games with tree search to pick a move at the end. To make this work, you need to update visit counts and Q-values at the end of the simulations. Visit counts are easy; you increment the count of a node by 1 if it has been traversed by search. To update Q-values, you sum up $V(l)$ for all visited leaf nodes *l* and divide by visit counts:

$$Q(s,a) = \sum_{i=1}^{n} \frac{V(l_i)}{N(s,a)}$$

Here you sum up over all *n* simulations and add the leaf node value of the *i*th simulation, if that simulation traversed the node corresponding to (*s,a*). To summarize this whole process, let's look at how you modified the four-step tree-search process from chapter 4:

1. *Select*—You traverse the game tree by choosing actions that maximize $Q(s,a) + u(s,a)$.
2. *Expand*—When expanding a new leaf, you ask the strong policy network once for prior probabilities to be stored for each child.
3. *Evaluate*—At the end of a simulation, a leaf node is evaluated by averaging the output of the value network with the outcome of a rollout that uses the fast policy.
4. *Update*—After all simulations are completed, you update visit counts and Q-values traversed in simulations.

The one thing we haven't discussed yet is how to pick a move to *play* after the simulations have finished. That's simple: you pick the most visited node! This might even seem a little too simplistic, but keep in mind that nodes get more and more visits over time as their Q-values improve. After you go through enough simulations, the node visit count will give you a great indication of a move's relative worth.

13.4.3 *Implementing AlphaGo's search algorithm*

Having discussed how AlphaGo uses neural networks in combination with tree search, let's go ahead and implement this algorithm in Python. Your goal is to create an `Agent` that has a `select_move` method as specified by the AlphaGo methodology. The code for this section can be found under dlgo/agent/alphago.py on GitHub.

You start with the full definition of an AlphaGo tree node, which you already sketched in the preceding section. An `AlphaGoNode` has a parent and children represented as a dictionary of moves to other nodes. A node also comes with a `visit _count`, a `q_value`, and a `prior_value`. Additionally, you store the *utility function* `u_value` of this node.

Listing 13.14   Defining an AlphaGo tree node in Python

```
import numpy as np
from dlgo.agent.base import Agent
from dlgo.goboard_fast import Move
```

```
from dlgo import kerasutil
import operator

class AlphaGoNode:
    def __init__(self, parent=None, probability=1.0):
        self.parent = parent
        self.children = {}

        self.visit_count = 0
        self.q_value = 0
        self.prior_value = probability
        self.u_value = probability
```

Tree nodes have one parent and potentially many children.

A node is initialized with a prior probability.

The utility function will be updated during search.

Such a node will be used by the tree-search algorithm in three places:

1. select_child—Traversing the tree in a simulation, you select children of a node according to argmax$_a Q(s,a) + u(s,a)$; you pick the action maximizing the sum of Q-value and utility function.

2. expand_children—At a leaf node, you'll ask the strong policy to evaluate all legal moves from this position and add new AlphaGoNode instances for each one of them.

3. update_values—Finally, after completing all simulations, you update visit _count, q_value, and u_value accordingly.

The first two of these methods are straightforward, as shown in the following listing.

Listing 13.15   Selecting an AlphaGo child by maximizing Q-value

```
class AlphaGoNode():
...

    def select_child(self):
        return max(self.children.items(),
                   key=lambda child: child[1].q_value + \
                   child[1].u_value)

    def expand_children(self, moves, probabilities):
        for move, prob in zip(moves, probabilities):
            if move not in self.children:
                self.children[move] = AlphaGoNode(probability=prob)
```

The third method to update the summary statistics of the AlphaGo node is a little more intricate. First of all, you use a slightly more sophisticated version of the utility function:

$$u(s,a) = c_u \sqrt{N_p(s,a)} \frac{P(s,a)}{1 + N(s,a)}$$

Compared to the version introduced in the preceding section, this utility has two extra terms. The first term c_u, which we'll call c_u in code, scales utility by a fixed *constant* for all nodes, which we set to 5 by default. The second term further scales utility by the square root of the parent's visit count (you denote the parent of the node in question by Np). This leads to higher utility of nodes whose parents have been visited more frequently.

Listing 13.16 Updating visit counts, Q-value, and utility of an AlphaGo node

```
class AlphaGoNode():
    ...
```

Increment the visit count for this node.

```
    def update_values(self, leaf_value):
        if self.parent is not None:
            self.parent.update_values(leaf_value)

        self.visit_count += 1

        self.q_value += leaf_value / self.visit_count
```

You update parents first to ensure that you traverse the tree from top to bottom.

Add the specified leaf value to the Q-value, normalized by visit count.

Update utility with current visit counts.

```
        if self.parent is not None:
            c_u = 5
            self.u_value = c_u * np.sqrt(self.parent.visit_count) \
                * self.prior_value / (1 + self.visit_count)
```

This completes the definition of `AlphaGoNode`. You can now use this tree structure in the search algorithm used in AlphaGo. The class `AlphaGoMCTS` that you'll be implementing is an `Agent` and is initialized with multiple arguments. First, you provide this agent with a fast and strong policy and a value network. Second, you need to specify AlphaGo-specific parameters for rollouts and evaluation:

- `lambda_value`—This is the λ value you use to weigh rollouts and value function against each other: $V(l) = \lambda \cdot \text{value}(l) + (1 - \lambda) \cdot \text{rollout}(l)$.
- `num_simulations`—This value specifies how many simulations will be run in the selection process of a move.
- `depth`—With this parameter, you tell the algorithm how many moves per simulation to look ahead (you specify the search depth).
- `rollout_limit`—When determining a leaf value, you run a policy rollout rollout(l). You use the parameter `rollout_limit` to tell AlphaGo how many moves to roll out before judging the outcome.

Listing 13.17 Initializing an `AlphaGoMCTS` Go playing agent

```
class AlphaGoMCTS(Agent):
    def __init__(self, policy_agent, fast_policy_agent, value_agent,
                 lambda_value=0.5, num_simulations=1000,
                 depth=50, rollout_limit=100):
        self.policy = policy_agent
        self.rollout_policy = fast_policy_agent
        self.value = value_agent

        self.lambda_value = lambda_value
```

```
self.num_simulations = num_simulations
self.depth = depth
self.rollout_limit = rollout_limit
self.root = AlphaGoNode()
```

It's now time to implement the select_move method of this new Agent, which does pretty much all the heavy lifting in this algorithm. We sketched AlphaGo's tree-search procedure in the preceding section, but let's go through the steps one more time:

- When you want to play a move, the first thing you do is to run num_simulations simulations on your game tree.
- In each simulation, you carry out look-ahead search until the specified depth is reached.
- If a node doesn't have any children, *expand* the tree by adding new AlphaGoNode for each legal move, using the strong policy network for prior probabilities.
- If a node does have children, *select* one by choosing the move that maximizes Q-value plus utility.
- Play the move used in this simulation on the Go board.
- When the depth is reached, *evaluate* this leaf node by computing the combined value function from the value network and a policy rollout.
- Update all AlphaGo nodes with the leaf values from simulations.

This process is precisely what you'll implement in select_move. Note that this method uses two other utility methods that we'll discuss later: policy_probabilities and policy_rollout.

Listing 13.18 The main method in AlphaGo's tree-search process

```
class AlphaGoMCTS(Agent):
...

    def select_move(self, game_state):
        for simulation in range(self.num_simulations):      ←─┐ From the current state,
            current_state = game_state                          play out a number of
            node = self.root                                    simulations.
            for depth in range(self.depth):                 ←─┐ Play moves until the
                if not node.children:                           specified depth is reached.
                    if current_state.is_over():
                        break
                    moves, probabilities =                  ←─┐ ...expand them with
                        self.policy_probabilities(current_state)  probabilities from the
                    node.expand_children(moves, probabilities)    strong policy.

                move, node = node.select_child()            ←─┐ Compute the
                current_state = current_state.apply_move(move)   output of the
                                                                 value network
                value = self.value.predict(current_state)        and a rollout by
                rollout = self.policy_rollout(current_state)      the fast policy.
```

If the current node doesn't have any children...

Play moves until the specified depth is reached.

...expand them with probabilities from the strong policy.

If there are children, you can select one and play the corresponding move.

Compute the output of the value network and a rollout by the fast policy.

Update values for this node in the backup phase.

```
weighted_value = (1 - self.lambda_value) * value + \
    self.lambda_value * rollout
node.update_values(weighted_value)
```

Determine the combined value function.

You might have noticed at this point that although you ran all simulations, you still haven't played any move. You do so by playing the most visited node, after which the only thing left to do is set a new root node accordingly and return the suggested move.

Listing 13.19 Selecting the most visited node and updating the tree's root node

```
class AlphaGoMCTS(Agent):
    ...
    def select_move(self, game_state):
        ...

        move = max(self.root.children, key=lambda move:
                self.root.children.get(move).visit_count)

        self.root = AlphaGoNode()
        if move in self.root.children:
            self.root = self.root.children[move]
            self.root.parent = None

        return move
```

Pick the most visited child of the root as the next move.

If the picked move is a child, set the new root to this child node.

This already completes the main process of AlphaGo's tree search, so let's have a look at the two utility methods we left out earlier. `policy_probabilities`, used in node expansion, computes predictions of the strong policy network, restricts these predictions to legal moves, and then normalizes the remaining predictions. The method returns both legal moves and their normalized policy network predictions.

Listing 13.20 Computing normalized strong policy values for legal moves on the board

```
class AlphaGoMCTS(Agent):
    ...

    def policy_probabilities(self, game_state):
        encoder = self.policy._encoder
        outputs = self.policy.predict(game_state)
        legal_moves = game_state.legal_moves()
        if not legal_moves:
            return [], []
        encoded_points = [encoder.encode_point(move.point) for move in
    legal_moves if move.point]
        legal_outputs = outputs[encoded_points]
        normalized_outputs = legal_outputs / np.sum(legal_outputs)
        return legal_moves, normalized_outputs
```

The last helper method you need is `policy_rollout` to compute the game result of a rollout using the fast policy. All this method does is *greedily* select the strongest move according to the fast policy until the rollout limit is reached, and then see who won. You return 1 if the player to move next won, –1 if the other player won, and 0 if no result has been reached.

Listing 13.21   Playing until the `rollout_limit` is reached

```
class AlphaGoMCTS(Agent):
...

    def policy_rollout(self, game_state):
        for step in range(self.rollout_limit):
            if game_state.is_over():
                break
            move_probabilities = self.rollout_policy.predict(game_state)
            encoder = self.rollout_policy.encoder
            valid_moves = [m for idx, m in enumerate(move_probabilities)
                            if Move(encoder.decode_point_index(idx)) in
 game_state.legal_moves()]
            max_index, max_value = max(enumerate(valid_moves),
 key=operator.itemgetter(1))
            max_point = encoder.decode_point_index(max_index)
            greedy_move = Move(max_point)
            if greedy_move in game_state.legal_moves():
                game_state = game_state.apply_move(greedy_move)

        next_player = game_state.next_player
        winner = game_state.winner()
        if winner is not None:
            return 1 if winner == next_player else -1
        else:
            return 0
```

With all the work you put into developing the `Agent` framework and implementing an AlphaGo agent, you can now use an `AlphaGoMCTS` instance to play games easily.

Listing 13.22   Initializing an AlphaGo agent with three deep neural networks

```
from dlgo.agent import load_prediction_agent, load_policy_agent, AlphaGoMCTS
from dlgo.rl import load_value_agent
import h5py

fast_policy = load_prediction_agent(h5py.File('alphago_sl_policy.h5', 'r'))
strong_policy = load_policy_agent(h5py.File('alphago_rl_policy.h5', 'r'))
value = load_value_agent(h5py.File('alphago_value.h5', 'r'))

alphago = AlphaGoMCTS(strong_policy, fast_policy, value)
```

This agent can be used in the exact same way as all the other agents you've developed in chapters 7 to 12. In particular, you can register HTTP and GTP frontends for this agent, as you did in chapter 8. This makes it possible to play against your AlphaGo bot, let other bots play against it, or even register and run it on an online Go server (such as OGS, as shown in appendix E).

13.5 Practical considerations for training your own AlphaGo

In the preceding section, you developed a rudimentary version of the tree-search algorithm used in AlphaGo. This algorithm *can* lead to a superhuman level of Go game play, but you need to read the fine print to get there. You need to not only ensure that you do a good job at training all three deep neural networks used in AlphaGo, but also ensure that the simulations in tree search are run fast enough, so you don't have to wait hours on end for AlphaGo to suggest the next move. Here are a few pointers for you to make the most of it:

- The first step of training, supervised learning of policy networks, was run on a corpus of 160,000 games from KGS, translating into about 30 million game states. In total, DeepMind's AlphaGo team computed 340 million training steps.

- The good news is that you have access to the exact same data set; DeepMind used the KGS training set we introduced in chapter 7 as well. In principle, nothing stops you from running the same number of training steps. The bad news is that even if you have a state-of-the-art GPU, the training process may take many months, if not years.

- The AlphaGo team addressed this issue by *distributing* training across 50 GPUs, reducing training time to three weeks. This is unlikely an option for you, particularly because we haven't discussed how to train deep networks in a distributed way.

- What you can do to come up with satisfying results is scale down each part of the equation. Use one of the board encoders from chapter 7 or 8 and use much smaller networks than the AlphaGo policy and value networks introduced in this chapter. Also, start with a small training set first, so you get a feeling for the training process.

- In self-play, DeepMind generated 30 million distinct positions. This is vastly more than you can realistically hope to create. As a rule of thumb, try to generate as many self-play positions as human game positions from supervised learning.

- If you simply take the large networks laid out in this chapter and train them on very little data, you'll likely end up worse than running a smaller network on more data.

- The fast policy network is used frequently in rollouts, so to speed up tree search, make sure your fast policy is really small in the beginning, such as the networks you used in chapter 6.

- The tree-search algorithm you implemented in the preceding section computes simulations *sequentially*. To speed up the process, DeepMind *parallelized* search and used a total of 40 search threads. In this parallel version, multiple GPUs were used to evaluate deep networks in parallel, and multiple CPUs were used to carry out the other parts of the tree search.

- Running tree search on multiple CPUs is feasible in principle (recall that you used multithreading for data preparation in chapter 7 as well), but is a little too involved to cover here.

- What you can do to improve the game-play experience, trading speed for strength, is to reduce the number of simulations run and the search depth used in search. This won't lead to superhuman performance, but at least the system becomes playable.

As you can see from these points, although the method of combining supervised and reinforcement learning with tree search in this novel way was an impressive feat, the engineering effort that went into scaling network training, evaluation, and tree search deserves its fair share of the credit in building the first Go bot to play better than top professionals.

In the last chapter, you'll see the next development stage of the AlphaGo system. It not only skips the supervised learning from human game records, but *plays even stronger* than the original AlphaGo system implemented in this chapter.

13.6   *Summary*

- To power an AlphaGo system, you have to train three deep neural networks: two policy networks and a value network.

- The fast policy network is trained from human game-play data and has to be quick enough to run many rollouts in AlphaGo's tree-search algorithm. The outcome of rollouts is used to evaluate leaf positions.

- The strong policy network is first trained on human data and then improved with self-play, using the policy gradient algorithm. You use this network in AlphaGo to compute prior probabilities for node selection.

- The value network is trained on experience data generated from self-play and used for position evaluation for leaf nodes, together with policy rollouts.

- Selecting a move with AlphaGo means generating numerous simulations, traversing the game tree. After the simulation step is completed, the most visited node is selected.

- In the simulation, nodes are *selected* by maximizing Q-values plus utility.

- When a leaf is reached, a node gets *expanded* by using the strong policy for prior probabilities.

- A leaf node is evaluated by a combined value function, mixing the output of the value network with the outcome of a fast policy rollout.

- In the backup phase of the algorithm, visit counts, Q-values, and utility values receive updates according to the chosen actions.

14
AlphaGo Zero: Integrating tree search with reinforcement learning

This chapter covers

- Playing games with a variation on Monte Carlo tree search
- Integrating tree search into self-play for reinforcement learning
- Training a neural network to enhance a tree-search algorithm

After DeepMind revealed the second edition of AlphaGo, code-named *Master*, Go fans all over the world scrutinized its shocking style of play. Master's games were full of surprising new moves. Although Master was bootstrapped from human games, it was continuously enhanced with reinforcement learning, and that enabled it to discover new moves that humans didn't play.

This led to an obvious question: what if AlphaGo didn't rely on human games at all, but instead learned entirely using reinforcement learning? Could it still reach a superhuman level, or would it get stuck playing with beginners? Would it rediscover patterns played by human masters, or would it play in an incomprehensible new alien style? All these questions were answered when AlphaGo Zero (AGZ) was announced in 2017.

AlphaGo Zero was built on an improved reinforcement-learning system, and it trained itself from scratch without any input from human games. Although its first games were worse than any human beginner's, AGZ improved steadily and quickly surpassed every previous edition of AlphaGo.

To us, the most astonishing thing about AlphaGo Zero is how it does more with less. In many ways, AGZ is much simpler than the original AlphaGo. No more hand-crafted feature planes. No more human game records. No more Monte Carlo rollouts. Instead of two neural networks and three training processes, AlphaGo Zero used one neural network and one training process.

And yet AlphaGo Zero was stronger than the original AlphaGo! How is that possible?

First, AGZ used a truly massive neural network. The strongest version ran on a network with capacity roughly equivalent to 80 convolution layers—over four times the size of the original AlphaGo network.

Second, AGZ used an innovative new reinforcement-learning technique. The original AlphaGo trained its policy network alone, in a manner similar to what we described in chapter 10; later that policy network was used to improve tree search. In contrast, AGZ integrated tree search with reinforcement learning from the beginning. This algorithm is the focus of this chapter.

To start, we go over the structure of the neural network that AGZ trains. Next, we describe the tree-search algorithm in depth. AGZ uses the same tree search in both self-play and competitive games. After that, we cover how AGZ trains its network from its experience data. To wrap up, we briefly cover a few practical tricks that AGZ uses to make the training process stable and efficient.

14.1   *Building a neural network for tree search*

AlphaGo Zero uses a single neural network with one input and two outputs: one output produces a probability distribution over moves, and the other output produces a single value representing whether the game favors white or black. This is the same structure we used for actor-critic learning in chapter 12.

There's one small difference between the output of the AGZ network and the network we used in chapter 12, and the difference is around passing in the game. In previous cases where we implemented self-play, we hardcoded logic around passing to end the game. For example, the `PolicyAgent` self-play bot from chapter 9 included custom logic so that it wouldn't fill in its own eyes, thereby killing its own stones. If the only legal moves were self-destructive, the `PolicyAgent` would pass. This ensured that self-play games ended in a sensible place.

Because AGZ uses a tree search during self-play, you don't need that custom logic. You can treat a pass the same as any other move, and expect the bot to learn when passing is appropriate. If the tree search reveals that playing a stone will lose the game, it'll pick a pass instead. This means your action output needs to return a probability for passing along with every point on the board. Instead of returning a vector of size $19 \times 19 = 361$ to represent each point on the board, your network will produce a vector of size $19 \times 19 + 1 = 362$ to represent each point on the board *and* the pass move. Figure 14.1 illustrates this new move encoding.

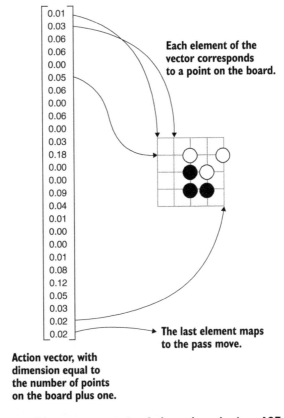

Figure 14.1 Encoding possible moves as a vector. As in previous chapters, AGZ uses a vector where each element maps to a point on the game board. AGZ adds one last element that maps to the pass move. This example is on a 5 × 5 board, so the vector has dimension 26: 25 for the points on the board and 1 for passing.

This means you have to modify the board encoder slightly. In previous board encoders, you implemented encode_point and decode_point_index, which translated between the elements of the vector and points on the board. For an AGZ-style bot, you'll replace these with new functions, encode_move and decode_move_index. The encoding for playing a stone remains the same; you use the next index to represent a pass.

Listing 14.1   Modifying the board encoder to include passing

```
class ZeroEncoder(Encoder):
...
    def encode_move(self, move):
        if move.is_play:
            return (self.board_size * (move.point.row - 1) +
                (move.point.col - 1))
        elif move.is_pass:
            return self.board_size * self.board_size
        raise ValueError('Cannot encode resign move')

    def decode_move_index(self, index):
        if index == self.board_size * self.board_size:
            return Move.pass_turn()
        row = index // self.board_size
        col = index % self.board_size
        return Move.play(Point(row=row + 1, col=col + 1))

    def num_moves(self):
        return self.board_size * self.board_size + 1
```

Same point encoding as used in previous encoders

Uses the next index to represent a pass

The neural network doesn't learn resignation.

Apart from the treatment of passing, the input and output of the AGZ network are identical to what we covered in chapter 12. For the inner layers of the network, AGZ used an extremely deep stack of convolution layers, with a few modern enhancements to make training smoother (we cover those briefly at the end of this chapter). A large network is powerful, but it also requires more computation, both for training and self-play. If you don't have access to the same kind of hardware as DeepMind, you may have better luck with a smaller network. Feel free to experiment to find the right balance of power and speed for your needs.

As for board encoding, you could use any encoding scheme we covered in this book, from the basic encoder in chapter 6 up to the 48-plane encoder in chapter 13. AlphaGo Zero used the simplest possible encoder: just the location of the black and white stones on the board, plus a plane indicating whose turn it is. (To handle ko, AGZ also included planes for the previous seven board positions.) But there's no technical reason why you can't use game-specific feature planes, and it's possible they'd make learning faster. In part, the researchers wanted to remove as much human knowledge as they could just to prove it was possible. In your own experiments, you should feel free to try different combinations of feature planes while using the AGZ reinforcement-learning algorithm.

14.2 Guiding tree search with a neural network

In reinforcement learning, a *policy* is an algorithm that tells an agent how to make decisions. In previous examples of reinforcement learning, the policy was relatively simple. In policy gradient learning (chapter 10) and actor-critic learning (chapter 12), a neural network directly told you which move to pick: that's the policy. In Q-learning (chapter 11), the policy involved computing the Q-value for each possible move; then you picked the move with the highest Q.

AGZ's policy includes a form of tree search. You'll still use a neural network, but the purpose of the neural network is to guide the tree search, rather than to choose or evaluate moves directly. Including tree search during self-play means the self-play games are more realistic. In turn, that means the training process is more stable.

The tree-search algorithm builds on ideas you've already studied. If you've studied the Monte Carlo tree-search algorithm (chapter 4) and the original AlphaGo (chapter 13), the AlphaGo Zero tree-search algorithm will seem familiar; table 14.1 compares the three algorithms. First, we'll describe the data structure that AGZ uses to represent a game tree. Next, we'll walk through the algorithm that AGZ uses to add a new position to the game tree.

Table 14.1 Comparing tree-search algorithms

	MCTS	AlphaGo	AlphaGo Zero
Branch selection	UCT score	UCT score + prior from policy network	UCT score + prior from combined network
Branch evaluation	Randomized playouts	Value network + randomized playouts	Value from combined network

The general idea of tree-search algorithms, as they apply to board games, is to find the move that leads to the best outcome. You determine that by examining possible sequences of moves that might follow. But the number of possible sequences is beyond enormous, so you need to make a decision while examining only a tiny fraction of the possible sequences. The art and science of tree-search algorithms is in how to pick the branches to explore in order to get the best result in the least time.

Just as in MCTS, the AGZ tree-search algorithm runs for a certain number of rounds, and in each round it adds another board position to the tree. As you execute more and more rounds, the tree continues to grow larger, and the algorithm's estimates get more accurate. For the purposes of illustration, imagine that you're already in the middle of the algorithm: you've already built up a partial tree, and you want to expand the tree with a new board position. Figure 14.2 shows such an example game tree.

Each node in the game tree represents a possible board position. From that position, you also know which follow-up moves are legal. The algorithm has already visited some of those follow-up moves, but not all of them. You create a *branch* for each follow-up move, whether you've visited or not. Each branch tracks the following:

- A *prior* probability of the move, indicating how good you expect this move to be, before you try visiting it.
- The number of times you've visited the branch during the tree search. This may be zero.
- The *expected value* of all visits that passed through this branch. This is an average over all visits through the tree. To make updating this average easier, you store the sum of the values; you can then divide by the visit count to get the average.

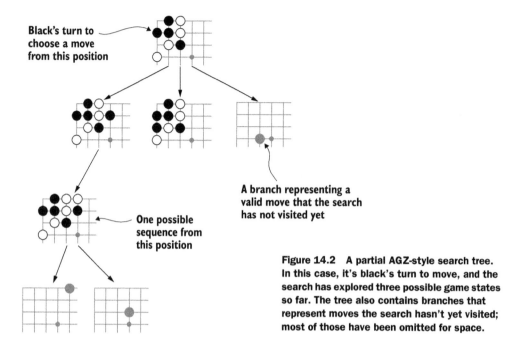

Black's turn to choose a move from this position

A branch representing a valid move that the search has not visited yet

One possible sequence from this position

Figure 14.2   A partial AGZ-style search tree. In this case, it's black's turn to move, and the search has explored three possible game states so far. The tree also contains branches that represent moves the search hasn't yet visited; most of those have been omitted for space.

For each branch that you *have* visited, the node also contains a pointer to a *child node*. In the following listing, you define a minimal `Branch` structure to contain branch statistics.

Listing 14.2   A structure to track branch statistics

```
class Branch:
    def __init__(self, prior):
        self.prior = prior
        self.visit_count = 0
        self.total_value = 0.0
```

Now you're ready to build the structure that represents the search tree. The following listing defines a `ZeroTreeNode` class.

Listing 14.3   A node in an AGZ-style search tree

```
class ZeroTreeNode:
    def __init__(self, state, value, priors, parent, last_move):
        self.state = state
        self.value = value
        self.parent = parent
        self.last_move = last_move
        self.total_visit_count = 1
        self.branches = {}
```

> In the root of the tree, parent and last_move will be None.

```
for move, p in priors.items():
    if state.is_valid_move(move):
        self.branches[move] = Branch(p)
self.children = {}
```

Later, children will map from a
Move to another ZeroTreeNode.

```
def moves(self):
    return self.branches.keys()
```

Returns a list of all possible
moves from this node

```
def add_child(self, move, child_node):
    self.children[move] = child_node
```

Allows adding new
nodes into the tree

Returns a
particular
child node

```
def has_child(self, move):
    return move in self.children
```

Checks whether there's a child
node for a particular move

The ZeroTreeNode class also includes some helpers for reading the statistics off its
children.

Listing 14.4   Helpers to read branch information from a tree node

```
class ZeroTreeNode:
...
    def expected_value(self, move):
        branch = self.branches[move]
        if branch.visit_count == 0:
            return 0.0
        return branch.total_value / branch.visit_count

    def prior(self, move):
        return self.branches[move].prior

    def visit_count(self, move):
        if move in self.branches:
            return self.branches[move].visit_count
        return 0
```

14.2.1 *Walking down the tree*

In each round of the search, you start by walking down the tree. The point is to see
what a possible future board position is, so you can evaluate whether it's good. To get
an accurate evaluation, you should assume that your opponent will respond to your
moves in the strongest possible way. Of course, you don't know what the strongest
response is yet; you must try out a variety of moves to find out which are good. This
section describes an algorithm for selecting strong moves in the face of uncertainty.

The expected value provides an estimate of how good each possible move is. But
the estimates aren't equally accurate. If you've spent more time reading out a particular branch, its estimate will be better.

You can continue to read out one of the best variations in more detail, which will
further improve its estimate. Alternately, you can read out a branch that you've
explored less, in order to improve your estimate. Maybe that move is better than you

initially thought; the only way to find out is expanding it further. Once again, you see the opposing goals of *exploitation* and *exploration*.

The original MCTS algorithm used the UCT (upper confidence bound for trees; see chapter 4) formula to balance these goals. The UCT formula balanced two priorities:

- If you've visited a branch many times, you trust its expected value. In that case, you prefer the branches with higher estimated values.
- If you've visited a branch only a few times, its expected value may be way off. Whether its expected value is good or bad, you want to visit it a few times to improve its estimate.

AGZ adds a third factor:

- Among branches with few visits, you prefer the ones with a high prior. Those are the moves that intuitively look good, before considering the exact details of this game.

Mathematically, AGZ's scoring function looks like this:

$$Q + cP\frac{\sqrt{N}}{1+n}$$

The parts of the equation are as follows:

- Q is the expected value averaged over all visits through a branch. (It's zero if you haven't visited the branch yet.)
- P is the prior for the move under consideration.
- N is the number of visits to the *parent* node.
- n is the number of visits to the *child* branch.
- c is a factor that balances exploration against exploitation (generally, you have to set this by trial and error).

Look at the example in figure 14.3. Branch A has been visited twice and has a slightly good evaluation of $Q = 0.1$. Branch B has been visited once and has a bad evaluation: $Q = -0.5$. Branch C has no visits yet but has a prior probability of $P = 0.038$.

Table 14.2 shows how to calculate the uncertainty component. Branch A has the highest Q component, indicating that you've seen some good board positions underneath it. Branch C has the highest UCT component: we've never visited, so we have the highest uncertainty around that branch. Branch B has a lower evaluation than A, and more visits than C, so it's not likely to be a good choice at this point.

Table 14.2 Choosing a branch to follow

	Q	n	N	P	P√N / (n + 1)
Branch A	0.1	2	3	0.068	0.039
Branch B	–0.5	1	3	0.042	0.036
Branch C	0	0	3	0.038	0.065

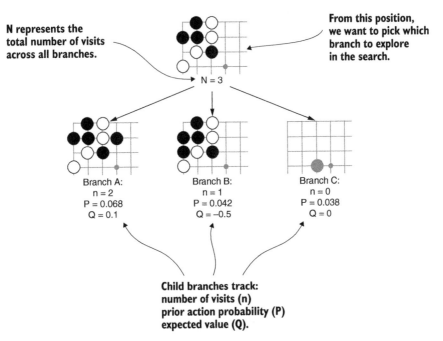

N represents the total number of visits across all branches.

From this position, we want to pick which branch to explore in the search.

N = 3

Branch A:
n = 2
P = 0.068
Q = 0.1

Branch B:
n = 1
P = 0.042
Q = –0.5

Branch C:
n = 0
P = 0.038
Q = 0

Child branches track: number of visits (n) prior action probability (P) expected value (Q).

Figure 14.3 Choosing which branch to follow in the AGZ tree search. In this example, you're considering three branches from the starting position. (In reality, there would be many more possible moves, which we omitted for space.) To choose a branch, you consider the number of times you've already visited the branch, your estimated value for the branch, and the prior probability of the move.

Assuming you've eliminated branch B, how do you choose between branch A and branch C? It depends on the value of the parameter c. A small value of c favors the high-value branch (in this case, A). A large value of c favors the branch with most uncertainty (in this case, C). For example, at $c = 1.0$, you'd choose A (at a score of 0.139 to 0.065). At $c = 4.0$, you'd choose C (0.260 to 0.256). Neither is objectively right; it's just a trade-off. The following listing shows how to calculate this score in Python.

Listing 14.5 Choosing a child branch

```
class ZeroAgent(Agent):
...
    def select_branch(self, node):
        total_n = node.total_visit_count

        def score_branch(move):
            q = node.expected_value(move)
            p = node.prior(move)
            n = node.visit_count(move)
            return q + self.c * p * np.sqrt(total_n) / (n + 1)

        return max(node.moves(), key=score_branch)
```

node.moves() is a list of moves. When you pass in key=score_branch, then max will return the move with the highest value of the score_branch function.

After you've chosen a branch, you repeat the same calculation on its children to choose the next branch. You continue the same process until you reach a branch with no children.

Listing 14.6   Walking down the search tree

```
class ZeroAgent(Agent):
    ...
    def select_move(self, game_state):
        root = self.create_node(game_state)

        for i in range(self.num_rounds):
            node = root
            next_move = self.select_branch(node)
            while node.has_child(next_move):
                node = node.get_child(next_move)
                next_move = self.select_branch(node)
```

> The next section shows the implementation of create_node.

> **When has_child returns False, you've reached the bottom of the tree.**

> This is the first step in a process that repeats many times per move. self.num_moves controls the number of times you repeat the search process.

14.2.2 Expanding the tree

At this point, you've reached an unexpanded branch of the tree. You can't search any further, because there's no node in the tree corresponding to the current move. The next step is to create a new node and add it to the tree.

To create a new node, you take the previous game state and apply the current move to get a new game state. You can then feed the new game state to your neural network, which gives you two valuable things. First, you get the prior estimates for all possible follow-up moves from the new game state. Second, you get the estimated value of the new game state. You use this information to initialize the statistics for the branches from this new node.

Listing 14.7   Creating a new node in the search tree

```
class ZeroAgent(Agent):
    ...
    def create_node(self, game_state, move=None, parent=None):
        state_tensor = self.encoder.encode(game_state)
        model_input = np.array([state_tensor])
        priors, values = self.model.predict(model_input)
        priors = priors[0]
        value = values[0][0]
        move_priors = {
            self.encoder.decode_move_index(idx): p
            for idx, p in enumerate(priors)
        }
        new_node = ZeroTreeNode(
            game_state, value,
            move_priors,
            parent, move)
        if parent is not None:
            parent.add_child(move, new_node)
        return new_node
```

> The Keras predict function is a batch function that takes an array of examples. You must wrap your board_tensor in an array.

> **Likewise, predict returns arrays with multiple results, so you must pull out the first item.**

> Unpack the priors vector into a dictionary mapping from Move objects to their corresponding prior probabilities.

Finally, you walk back up the tree and update the statistics for each parent that lead to this node, as shown in figure 14.4. For each node in this path, you increment the visit count and update the total expected value. At each node, the perspective switches between the black player and the white player, so you need to flip the sign of the value at each step.

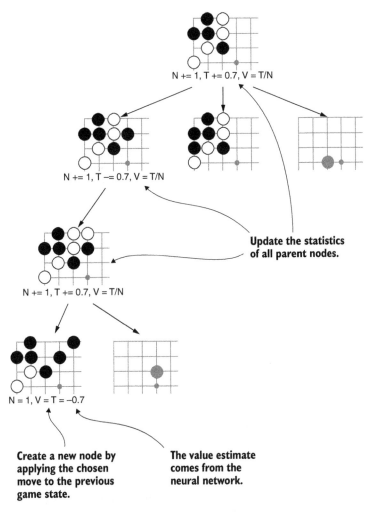

Update the statistics
of all parent nodes.

Create a new node by
applying the chosen
move to the previous
game state.

The value estimate
comes from the
neural network.

Figure 14.4 Expanding an AGZ-style search tree. First, you calculate a new game state. You create a new node from that game state and add it to the tree. The neural network then gives you a value estimate for that game state. Finally, you update all parents of the new node. You increment the visit count N by one and update the average value V. Here, T represents the total value across all visits through a node; that's just bookkeeping to make it easy to recalculate the average.

Listing 14.8   Expanding the search tree and updating all node statistics

```
class ZeroTreeNode:
    ...
    def record_visit(self, move, value):
        self.total_visit_count += 1
        self.branches[move].visit_count += 1
        self.branches[move].total_value += value

class ZeroAgent(Agent):
    ...
    def select_move(self, game_state):
    ...
            new_state = node.state.apply_move(next_move)
            child_node = self.create_node(
                new_state, parent=node)

            move = next_move
            value = -1 * child_node.value        ◄─────┐  At each level in the tree, you
            while node is not None:                     │  switch perspective between the
                node.record_visit(move, value)          │  two players. Therefore, you must
                move = node.last_move                    │  multiply the value by −1: what's
                node = node.parent                       │  good for black is bad for white,
                value = -1 * value                       │  and vice versa.
```

This whole process repeats over and over, and the tree expands each time. AGZ used 1,600 rounds per move during the self-play process. In competitive games, you should run the algorithm for as many rounds as you have time for. The bot will choose better and better moves as it performs more rounds.

14.2.3 Selecting a move

After you've built up the search tree as deeply as you can, it's time to select a move. The simplest rule for move selection is to pick the move with the highest visit count.

Why use the visit counts and not the expected value? You can assume the branch with the most visits has a high expected value. Here's why. Refer to the preceding branch-selection formula. As the number of visits to a branch grows, the factor of $1 / (n + 1)$ gets smaller and smaller. Therefore, the branch selection function will just pick based on Q. The branches with higher Q values get the most visits.

Now, if a branch has only a few visits, anything's possible. It may have a small Q or a huge Q. But you also can't trust its estimate based on a small number of visits. If you just picked the branch with the highest Q, you might get a branch with just one visit, and its true value may be much smaller. That's why you select based on visit counts instead. It guarantees that you pick a branch with a high estimated value *and* a reliable estimate.

Listing 14.9   Selecting the move with the highest visit count

```
class ZeroAgent(Agent):
    ...
    def select_move(self, game_state):
    ...
        return max(root.moves(), key=root.visit_count)
```

In contrast to other self-play agents in this book, the ZeroAgent has no special logic around when to pass. That's because you include passing in the search tree: you can treat it just like any other move.

With our ZeroAgent implementation complete, you can now implement your simulate_game function for self-play.

Listing 14.10 Simulating a self-play game

```
def simulate_game(
        board_size,
        black_agent, black_collector,
        white_agent, white_collector):
    print('Starting the game!')
    game = GameState.new_game(board_size)
    agents = {
        Player.black: black_agent,
        Player.white: white_agent,
    }

    black_collector.begin_episode()
    white_collector.begin_episode()
    while not game.is_over():
        next_move = agents[game.next_player].select_move(game)
        game = game.apply_move(next_move)

    game_result = scoring.compute_game_result(game)
    if game_result.winner == Player.black:
        black_collector.complete_episode(1)
        white_collector.complete_episode(-1)
    else:
        black_collector.complete_episode(-1)
        white_collector.complete_episode(1)
```

14.3 Training

The training target for the value output is a 1 if the agent won the game, and a –1 if it lost. By averaging over many games, you learn a value between those extremes that indicates your bot's chance of winning. It's the exact same setup you used for Q-learning (in chapter 11) and for actor-critic learning (chapter 12).

The action output is a little different. Just as in policy learning (chapter 10) and actor-critic learning (chapter 12), the neural network has an output that produces a probability distribution over legal moves. In policy learning, you trained a network to match the exact move that the agent chose (in cases where the agent won the game). AGZ works differently in a subtle way. It trains its network to match the number of times it visited each move during the tree search.

To illustrate how that can improve its playing strength, think about how the MCTS-style search algorithms work. Assume, for the moment, that you have a value function that's at least vaguely correct; it doesn't have to be super precise as long as it roughly separates winning positions from losing positions. Then imagine that you throw out the prior probabilities entirely and run the search algorithm. By design, the search will spend more time in the most promising branches. The branch selection logic makes this happen: the Q component in the UCT formula means high-value branches are selected more often. If you had unlimited time to run the search, it'd eventually converge on the best moves.

After a sufficient number of rounds in the tree search, you can think of the visit counts as the source of truth. You know whether these moves are good or bad because you checked what happens if you play them. So the search counts become your target values for training the prior function.

The prior function tries to predict where the tree search would spend its time, if you gave it plenty of time to run. Armed with a function trained on previous runs, your tree search can save time and go straight into searching out the more important branches. With an accurate prior function, your search algorithm can spend just a small number of rollouts, but get similar results to a slower search that required a much larger number of rollouts. In a sense, you can think that the network is "remembering" what happened in previous searches and using that knowledge to skip ahead.

To set up training in this way, you need to store the search counts after each move. In previous chapters, you used a generic `ExperienceCollector` that could apply to many RL implementations. In this case, however, the search counts are specific to AGZ, so you'll make a custom collector. The structure is mostly the same.

Listing 14.11   A specialized experience collector for AGZ-style learning

```python
class ZeroExperienceCollector:
    def __init__(self):
        self.states = []
        self.visit_counts = []
        self.rewards = []
        self._current_episode_states = []
        self._current_episode_visit_counts = []

    def begin_episode(self):
        self._current_episode_states = []
        self._current_episode_visit_counts = []

    def record_decision(self, state, visit_counts):
        self._current_episode_states.append(state)
        self._current_episode_visit_counts.append(visit_counts)

    def complete_episode(self, reward):
        num_states = len(self._current_episode_states)
        self.states += self._current_episode_states
        self.visit_counts += self._current_episode_visit_counts
        self.rewards += [reward for _ in range(num_states)]

        self._current_episode_states = []
        self._current_episode_visit_counts = []
```

```
Listing 14.12   Passing along the decision to the experience collector

class ZeroAgent(Agent):
...
    def select_move(self, game_state):
...
        if self.collector is not None:
            root_state_tensor = self.encoder.encode(game_state)
            visit_counts = np.array([
                root.visit_count(
                    self.encoder.decode_move_index(idx))
                    for idx in range(self.encoder.num_moves())
            ])
            self.collector.record_decision(
                root_state_tensor, visit_counts)
```

The action output of your neural network uses a softmax activation. Recall that the softmax activation ensures that its values sum to 1. For training, you should also make sure that the training target sums to 1. To do this, divide the total visit counts by its sum; this operation is called *normalizing*. Figure 14.5 shows an example.

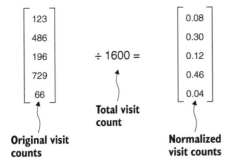

Figure 14.5   Normalizing a vector. During self-play, you track the number of times you visit each move. For training, you must normalize the vector so it sums to 1.

Other than that, the training process looks similar to training an actor-critic network in chapter 12. The following listing shows the implementation.

```
Listing 14.13   Training the combined network
```

Normalizes the visit counts. When you call np.sum with
axis=1, it sums along each row of the matrix. The
reshape call reorganizes those sums into matching rows.
Then you can divide the original counts by their sums.

```
class ZeroAgent(Agent):
...
    def train(self, experience, learning_rate, batch_size):
        num_examples = experience.states.shape[0]

        model_input = experience.states

        visit_sums = np.sum(
            experience.visit_counts, axis=1).reshape(
            (num_examples, 1))
        action_target = experience.visit_counts / visit_sums
```

See chapter 10 for a discussion of learning_rate and batch_size.

```
        value_target = experience.rewards

        self.model.compile(
            SGD(lr=learning_rate),
            loss=['categorical_crossentropy', 'mse'])
        self.model.fit(
            model_input, [action_target, value_target],
            batch_size=batch_size)
```

The overall reinforcement-learning cycle is the same as what you studied in chapters 9 through 12:

1 Generate a huge batch of self-play games.
2 Train the model on the experience data.
3 Test the updated model against the previous version.
4 If the new version is measurably stronger, switch to the new version.
5 If not, generate more self-play games and try again.
6 Repeat as many times as needed.

Listing 14.14 shows an example that runs a single cycle of this process. Fair warning: you'll need a *lot* of self-play games to build a strong Go AI from nothing. AlphaGo Zero reached a superhuman level of play, but it took nearly 5 million self-play games to get there.

> **Listing 14.14   A single cycle of the reinforcement-learning process**

```
board_size = 9
encoder = zero.ZeroEncoder(board_size)

board_input = Input(shape=encoder.shape(), name='board_input')
pb = board_input
for i in range(4):
    pb = Conv2D(64, (3, 3),
        padding='same',
        data_format='channels_first',
        activation='relu')(pb)
```
Build a network with four convolutional layers. To build a strong bot, you can add many more layers.

```
policy_conv = \
    Conv2D(2, (1, 1),
        data_format='channels_first',
        activation='relu')(pb)
policy_flat = Flatten()(policy_conv)
policy_output = \
    Dense(encoder.num_moves(), activation='softmax')(
        policy_flat)
```
Add the action output to the network.

```
value_conv = \
    Conv2D(1, (1, 1),
        data_format='channels_first',
        activation='relu')(pb)
```
Add the value output to the network.

```
value_flat = Flatten()(value_conv)
value_hidden = Dense(256, activation='relu')(value_flat)
value_output = Dense(1, activation='tanh')(value_hidden)

model = Model(
    inputs=[board_input],
    outputs=[policy_output, value_output])

black_agent = zero.ZeroAgent(
    model, encoder, rounds_per_move=10, c=2.0)
white_agent = zero.ZeroAgent(
    model, encoder, rounds_per_move=10, c=2.0)
c1 = zero.ZeroExperienceCollector()
c2 = zero.ZeroExperienceCollector()
black_agent.set_collector(c1)
white_agent.set_collector(c2)

for i in range(5):
    simulate_game(board_size, black_agent, c1, white_agent, c2)

exp = zero.combine_experience([c1, c2])
black_agent.train(exp, 0.01, 2048)
```

> **Add the value output to the network.**

> **We use 10 rounds per move here just so the demo runs quickly. For real training, you'll need much more; AGZ used 1,600.**

> **Simulate five games before training. For real training, you'll want to train on much larger batches (thousands of games).**

14.4 *Improving exploration with Dirichlet noise*

Self-play reinforcement learning is an inherently random process. Your bot can easily drift in a weird direction, especially early in the training process. To prevent the bot from getting stuck, it's important to provide a little randomness. That way, if the bot gets fixated on a really terrible move, it'll have a small chance to learn a better move. In this section, we describe one of the tricks AGZ used to ensure good exploration.

In previous chapters, you used a few different techniques to add variety to a bot's selections. For example, in chapter 9, you randomly sampled from your bot's policy output; in chapter 11, you used the ε-greedy algorithm: some fraction ε of the time, the bot would ignore its model entirely and choose a totally random move instead. In both cases, you added randomness at the time the bot made a decision. AGZ uses a different method to introduce randomness earlier, during the search process.

Imagine that on each turn, you artificially boosted the prior of one or two randomly chosen moves. Early in the search process, the prior controls which branches get explored, so those moves will get extra visits. If they turn out to be bad moves, the search will quickly move on to other branches, so no harm done. But this would ensure that every move gets a few visits occasionally, so the search doesn't develop blind spots.

AGZ achieves a similar effect by adding noise—small random numbers—to the priors at the root of each search tree. By drawing the noise from a *Dirichlet distribution*, you get the exact effect described previously: a few moves get an artificial boost, while the others stay untouched. In this section, we explain the properties of the Dirichlet distribution and show how to generate Dirichlet noise by using NumPy.

Throughout this book, you've used probability distributions over game moves. When you sample from such a distribution, you get a particular move. The Dirichlet distribution is a probability distribution over probability distributions: when you sample from the Dirichlet distribution, you get another probability distribution. The NumPy function `np.random.dirichlet` generates samples from a Dirichlet distribution. It takes a vector argument and returns a vector of the same dimension. The following listing shows a few example draws: the result is a vector, and it always sums to 1—meaning the result is itself a valid probability distribution.

Listing 14.15   Using `np.random.dirichlet` to sample from a Dirichlet distribution

```
>>> import numpy as np
>>> np.random.dirichlet([1, 1, 1])
array([0.1146, 0.2526, 0.6328])
>>> np.random.dirichlet([1, 1, 1])
array([0.1671, 0.5378, 0.2951])
>>> np.random.dirichlet([1, 1, 1])
array([0.4098, 0.1587, 0.4315])
```

You can control the output of the Dirichlet distribution with a *concentration* parameter, usually denoted as α. When α is close to 0, the Dirichlet distribution will generate "lumpy" vectors: most of the values with be close to 0, and just a few values will be larger. When α is large, the samples will be "smooth": the values will be closer to each other. The following listing shows the effect of changing the concentration parameter.

Listing 14.16   Drawing from a Dirichlet distribution when α is close to zero

```
>>> import numpy as np

>>> np.random.dirichlet([0.1, 0.1, 0.1, 0.1])
array([0.      , 0.044 , 0.7196, 0.2364])
>>> np.random.dirichlet([0.1, 0.1, 0.1, 0.1])
array([0.0015, 0.0028, 0.9957, 0.      ])
>>> np.random.dirichlet([0.1, 0.1, 0.1, 0.1])
array([0.      , 0.9236, 0.0002, 0.0763])

>>> np.random.dirichlet([10, 10, 10, 10])
array([0.3479, 0.1569, 0.3109, 0.1842])
>>> np.random.dirichlet([10, 10, 10, 10])
array([0.3731, 0.2048, 0.0715, 0.3507])
>>> np.random.dirichlet([10, 10, 10, 10])
array([0.2119, 0.2174, 0.3042, 0.2665])
```

Drawing from a Dirichlet distribution with a small concentration parameter. The results are "lumpy": most of the mass is concentrated in one or two elements.

Drawing from a Dirichlet distribution with a large concentration parameter. In each result, the mass is spread evenly over all elements of the vector.

This shows a recipe for modifying your priors. By choosing a small α, you get a distribution in which a few moves have high probabilities, and the rest are close to zero. Then you can take a weighted average of the true priors with the Dirichlet noise. AGZ used a concentration parameter of 0.03.

14.5 Modern techniques for deeper neural networks

Neural network design is a hot research topic. One never-ending problem is how to make training stable on deeper and deeper networks. AlphaGo Zero applied a couple of cutting-edge techniques that are quickly becoming standards. The details are beyond the scope of this book, but we introduce them at a high level here.

14.5.1 Batch normalization

The idea of deep neural networks is that each layer can learn an increasingly high-level representation of the original data. But what exactly are these representations? What we mean is that some meaningful property of the original data will show up as a particular numeric value in the activation of a particular neuron in the layer. But the mapping between actual numbers is completely arbitrary. If you multiply every activation in a layer by 2, for example, you haven't lost any information: you've just changed the scale. In principle, such a transformation doesn't affect the network's capacity to learn.

But the absolute value of the activations can affect practical training performance. The idea of *batch normalization* is to shift each layer's activations so they're centered around 0, and scale them so the variance is 1. At the start of training, you don't know what the activations will look like. Batch normalization provides a scheme for learning the correct shift and scale on the fly during the training process; the normalizing transform will adjust as its inputs change throughout training.

How does batch normalization improve training? That's still an open area of research. The original researchers developed batch normalization in order to reduce *covariate shift*. The activations in any layer tend to drift during training. Batch normalization corrects that drift, reducing the learning burden on the later layers. But the latest research suggests that the covariate shift may not be as important as initially thought. Instead, the value may be in the way batch normalization makes the loss function smoother.

Although researchers are still studying *why* batch normalization works, it's well established that it *does* work. Keras provides a `BatchNormalization` layer that you can add to your networks. The following listing shows an example of adding batch normalization to a convolution layer in Keras.

Listing 14.17 Adding batch normalization to a Keras network

The axis should match the convolution data_format. For channels_first, use axis=1 (the first axis). For channels_last, use axis=-/1 (the last axis).

```
from keras.models import Sequential
from keras.layers import Activation, BatchNormalization, Conv2D

model = Sequential()
model.add(Conv2D(64, (3, 3), data_format='channels_first'))
model.add(BatchNormalization(axis=1))
model.add(Activation('relu'))
```

The normalization happens between the convolution and the relu activation.

14.5.2 Residual networks

Imagine you've successfully trained a neural network that has three hidden layers in the middle. What happens if you add a fourth layer? In theory, this should strictly increase the capacity of your network. In the worst case, when you train the four-layer network, the first three layers could learn the same things they did in the three-layer network, while the fourth layer just passes its input untouched. You'd hope it could also learn *more*, but you wouldn't expect it to learn *less*. At least, you'd expect the deeper network should be able to overfit (memorize the training set in a way that doesn't necessarily translate to new examples).

In reality, this doesn't always happen. When you try to train a four-layer network, there are more possible ways to organize the data than on a three-layer network. Sometimes, because of the quirks of stochastic gradient descent on complicated loss surfaces, you can add more layers and find you can't even overfit. The idea of a *residual network* is to simplify what the extra layer is trying to learn. If three layers can do an OK job learning a problem, you can force the fourth layer to focus on learning the gap between whatever the first three layers learned and the objective. (That gap is the *residual*, hence the name.)

To implement this, you sum the *input* to your extra layers with the *output* from your extra layers, as illustrated in figure 14.6. The connection from the previous layer to the summation layer is called a *skip connection*. Normally, residual networks are organized into small blocks; there are around two or three layers per block with a skip connection beside them. Then you can stack as many blocks together as needed.

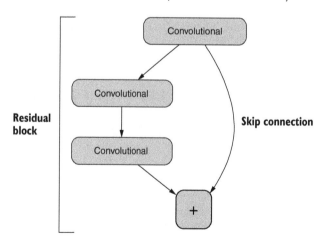

Figure 14.6   A residual block. The output of the two inner layers is added to the output of the previous layer. The effect is that the inner layers learn the difference, or residual, between the objective and what the previous layer learned.

14.6   Exploring additional resources

If you're interested in experimenting with AlphaGo Zero-style bots, there are many open source projects inspired by the original AGZ paper. If you want a superhuman Go AI, either to play against or to study the source code, you have an embarrassment of riches.

- Leela Zero is an open source implementation of an AGZ-style bot. The self-play process is distributed: if you have CPU cycles to spare, you can generate self-play games and upload them for training. At this writing, the community has contributed over 8 million games, and Leela Zero is already strong enough to beat professional Go players. http://zero.sjeng.org/

- Minigo is another open source implementation, written in Python with TensorFlow. It's fully integrated with Google Cloud Platform so you can use Google's public cloud to run experiments. https://github.com/tensorflow/minigo

- Facebook AI Research implemented the AGZ algorithm on top of its ELF reinforcement learning platform. The result, ELF OpenGo, is now freely available, and it's among the strongest Go AIs today. https://facebook.ai/developers/tools/elf

- Tencent has also implemented and trained an AGZ-style bot, which they have released as PhoenixGo. The bot is also known as BensonDarr on the Fox Go server, where it has beaten many of the world's top players. https://github.com/Tencent/PhoenixGo

- If Go isn't your thing, Leela Chess Zero is a fork of Leela Zero that has been adapted to learn chess instead. It's already at least as strong as human grandmasters, and chess fans have praised its exciting and creative play. https://github.com/LeelaChessZero/lczero

14.7 Wrapping up

That wraps up our introduction to the cutting-edge AI techniques that power modern Go AIs. We encourage you to take matters into your own hands from here: either experiment with your own Go bot, or try applying these modern techniques to other games.

But also think beyond games. When you read about the latest application of machine learning, you now have a framework for understanding what's happening. Think about the following:

- What is the model or neural network structure?
- What is the loss function or training objective?
- What is the training process?
- How are the inputs and outputs encoded?
- How can the model fit in with traditional algorithms or practical software applications?

We hope that we've inspired you to try your own experiments with deep learning, whether they're in games or another field.

14.8 Summary

- AlphaGo Zero uses a single neural network with two outputs. One output indicates which moves are important, and the other output indicates which player is ahead.

- The AlphaGo Zero tree-search algorithm is similar to Monte Carlo tree search, with two major differences. Instead of using random games to evaluate a position, it relies solely on a neural network. In addition, it uses a neural network to guide the search toward new branches.
- AlphaGo Zero's neural network is trained against the number of times it visited particular moves in the search process. In that way, it's specifically trained to enhance tree search, rather than to select moves directly.
- A *Dirichlet distribution* is a probability distribution over probability distributions. The concentration parameter controls how clumpy the resulting probability distributions are. AlphaGo Zero uses Dirichlet noise to add controlled randomness to its search process, to make sure all moves get explored occasionally.
- Batch normalization and residual networks are two modern techniques that help you train very deep neural networks.

appendix A
Mathematical foundations

You can't do machine learning without math. In particular, linear algebra and calculus are essential. The goal of this appendix is to provide enough mathematical background to help you understand the code samples in the book. We don't have nearly enough space to cover these massive topics thoroughly; if you want to understand these subjects better, we provide some suggestions for further reading.

If you're already familiar with advanced machine-learning techniques, you can safely skip this appendix altogether.

Further reading

In this book, we have room to cover only a few mathematical basics. If you're interested in learning more about the mathematical foundations of machine learning, here are some suggestions:

- For a thorough treatment of linear algebra, we suggest Sheldon Axler's *Linear Algebra Done Right* (Springer, 2015).
- For a complete and practical guide to calculus, including vector calculus, we like James Stewart's *Calculus: Early Transcendentals* (Cengage Learning, 2015).
- If you're serious about understanding the mathematical theory of how and why calculus works, it's hard to beat Walter Rudin's classic *Principles of Mathematical Analysis* (McGraw Hill, 1976).

Vectors, matrices, and beyond: a linear algebra primer

Linear algebra provides tools for handling arrays of data known as *vectors*, *matrices*, and *tensors*. You can represent all of these objects in Python with NumPy's `array` type.

Linear algebra is fundamental to machine learning. This section covers only the most basic operations, with a focus on how to implement them in NumPy.

Vectors: one-dimensional data

A *vector* is a one-dimensional array of numbers. The size of the array is the dimension of the vector. You use NumPy arrays to represent vectors in Python code.

> **NOTE**   This isn't the true mathematical definition of a vector, but for the purposes of our book, it's close enough.

You can convert a list of numbers into a NumPy array with the np.array function. The shape attribute lets you check the dimension:

```
>>> import numpy as np
>>> x = np.array([1, 2])
>>> x
array([1, 2])
>>> x.shape
(2,)
>>> y = np.array([3, 3.1, 3.2, 3.3])
>>> y
array([3. , 3.1, 3.2, 3.3])
>>> y.shape
(4,)
```

Note that shape is always a tuple; this is because arrays can be multidimensional, as you'll see in the next section.

You can access individual elements of a vector, just as if it were a Python list:

```
>>> x = np.array([5, 6, 7, 8])
>>> x[0]
5
>>> x[1]
6
```

Vectors support a few basic algebraic operations. You can add two vectors of the same dimension. The result is a third vector of the same dimension. Each element of the sum vector is the sum of the matching elements in the original vectors:

```
>>> x = np.array([1, 2, 3, 4])
>>> y = np.array([5, 6, 7, 8])
>>> x + y
array([ 6,  8, 10, 12])
```

Similarly, you can also multiply two vectors element-wise with the * operator. (Here, element-wise means you multiply each pair of corresponding elements separately.)

```
>>> x = np.array([1, 2, 3, 4])
>>> y = np.array([5, 6, 7, 8])
>>> x * y
array([ 5, 12, 21, 32])
```

The element-wise product is also called the *Hadamard product*.

You can also multiply a vector with a single float (or *scalar*). In this case, you multiply each value in the vector by the scalar:

```
>>> x = np.array([1, 2, 3, 4])
>>> 0.5 * x
array([0.5, 1. , 1.5, 2. ])
```

Vectors support a third kind of multiplication, the *dot product*, or *inner product*. To compute the dot product, you multiply each pair of corresponding elements and sum the results. So the dot product of two vectors is a single float. The NumPy function `np.dot` calculates the dot product. In Python 3.5 and later, the `@` operator does the same thing. (In this book, we use `np.dot`.)

```
>>> x = np.array([1, 2, 3, 4])
>>> y = np.array([4, 5, 6, 7])
>>> np.dot(x, y)
60
>>> x @ y
60
```

Matrices: two-dimensional data

A two-dimensional array of numbers is called a *matrix*. You can also represent matrices with NumPy arrays. In this case, if you pass a list of lists into `np.array`, you get a two-dimensional matrix back:

```
>>> x = np.array([
   [1, 2, 3],
   [4, 5, 6]
 ])
>>> x
array([[1, 2, 3],
       [4, 5, 6]])
>>> x.shape
(2, 3)
```

Note that the shape of a matrix is a two-element tuple: first is the number of rows, and second is the number of columns. You can access single elements with a double-subscript notation: first row, then column. Alternately, NumPy lets you pass in the indices in a `[row, column]` format. Both are equivalent:

```
>>> x = np.array([
   [1, 2, 3],
   [4, 5, 6]
 ])
>>> x[0][1]
2
>>> x[0, 1]
2
>>> x[1][0]
```

```
4
>>> x[1, 0]
4
```

You can also pull out a whole row from a matrix and get a vector:

```
>>> x = np.array([
  [1, 2, 3],
  [4, 5, 6]
 ])
>>> y = x[0]
>>> y
array([1, 2, 3])
>>> y.shape
(3,)
```

To pull out a column, you can use the funny-looking notation [:, n]. If it helps, think of : as Python's list-slicing operator; so [:, n] means "get me all rows, but only column *n*." Here's an example:

```
>>> x = np.array([
  [1, 2, 3],
  [4, 5, 6]
 ])
>>> z = x[:, 1]
>>> z
array([2, 5])
```

Just like vectors, matrices support element-wise addition, element-wise multiplication, and scalar multiplication:

```
>>> x = np.array([
  [1, 2, 3],
  [4, 5, 6]
 ])
>>> y = np.array([
  [3, 4, 5],
  [6, 7, 8]
 ])
>>> x + y
array([[ 4,  6,  8],
       [10, 12, 14]])
>>> x * y
array([[ 3,  8, 15],
       [24, 35, 48]])
>>> 0.5 * x
array([[0.5, 1. , 1.5],
       [2. , 2.5, 3. ]])
```

Rank 3 tensors

Go is played on a grid; so are chess, checkers, and a variety of other classic games. Any point on the grid can contain one of a variety of different game pieces. How do you represent the contents of the board as a mathematical object? One solution is to represent the board as a stack of matrices, and each matrix is the size of the game board.

Each individual matrix in the stack is called a *plane*, or *channel*. Each channel can represent a single type of piece that can be on the game board. In Go, you might have one channel for black stones and a second channel for white stones; figure A.1 shows an example. In chess, maybe you have a channel for pawns, another channel for bishops, one for knights, and so forth. You can represent the whole stack of matrices as a single three-dimensional array; this is called a *rank 3 tensor*.

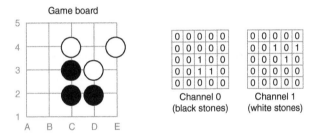

Figure A.1   Representing a Go game board with a two-plane tensor. This is a 5 × 5 board. You use one channel for black stones and a separate channel for white stones. So you use a 2 × 5 × 5 tensor to represent the board.

Another common case is representing an image. Let's say you want to represent a 128 × 64 pixel image with a NumPy array. In that case, you start with a grid corresponding to the pixels in the image. In computer graphics, you typically break a color into red, green, and blue components. So you can represent that image with a 3 × 128 × 64 tensor: you have a red channel, a green channel, and a blue channel.

Once again, you can use np.array to construct a tensor. The shape will be a tuple with three components, and you can use subscripting to pull out individual channels:

```
>>> x = np.array([
  [[1, 2, 3],
   [2, 3, 4]],
  [[3, 4, 5],
   [4, 5, 6]]
])
>>> x.shape
(2, 2, 3)
>>> x[0]
array([[1, 2, 3],
       [2, 3, 4]])
>>> x[1]
array([[3, 4, 5],
       [4, 5, 6]])
```

As with vectors and matrices, tensors support element-wise addition, element-wise multiplication, and scalar multiplication.

If you have an 8×8 grid with three channels, you could represent it with a $3 \times 8 \times 8$ tensor or an $8 \times 8 \times 3$ tensor. The only difference is in the way you index it. When you process the tensors with library functions, you must make sure the functions are aware of which indexing scheme you chose. The Keras library, which you use for designing neural networks, calls these two options `channels_first` and `channels_last`. For the most part, the choice doesn't matter: you just need to pick one and stick to it consistently. In this book, we use the `channels_first` format.

> **NOTE** If you need a motivation to pick a format, certain NVIDIA GPUs have special optimizations for the `channels_first` format.

Rank 4 tensors

In many places in the book, we use a rank 3 tensor to represent a game board. For efficiency, you may want to pass many game boards to a function at once. One solution is to pack the board tensors into a four-dimensional NumPy array: this is a tensor of rank 4. You can think of this four-dimensional array as a list of rank 3 tensors, each of which represents a single board.

Matrices and vectors are just special cases of tensors: a matrix is a rank 2 tensor, and a vector is a rank 1 tensor. And a rank 0 tensor is a plain old number.

Rank 4 tensors are the highest-order tensors you'll see in this book, but NumPy can handle tensors of any rank. Visualizing high-dimensional tensors is hard, but the algebra is the same.

Calculus in five minutes: derivatives and finding maxima

In calculus, the rate of change of a function is called its *derivative*. Table A.1 lists a few real-world examples.

Table A.1 Examples of derivatives

Quantity	Derivative
How far you've traveled	How fast you moved
How much water is in a tub	How fast the water drained out
How many customers you have	How many customers you gain (or lose)

A derivative is not a fixed quantity: it's another function that varies over time or space. On a trip in a car, you drive faster or slower at various times. But your speed is always connected to the distance you cover. If you had a precise record of where you were over time, you could go back and work out how fast you were traveling at any point in the trip. That is the derivative.

When a function is increasing, its derivative is positive. When a function is decreasing, its derivative is negative. Figure A.2 illustrates this concept. With this knowledge, you can use the derivative to find a *local maximum* or a *local minimum*. Any place the derivative is positive, you can move to the right a little bit and find a larger value. If you go past the maximum, the function must now be decreasing, and its derivative is negative. In that case, you want to move a little bit to the left. At the local maximum, the derivative will be exactly zero. The logic for finding a local minimum is identical, except you move in the opposite direction.

Many functions that show up in machine learning take a high-dimensional vector as input and compute a single number as output. You can extend the same idea to maximize or minimize such a function. The derivative of such a function is a vector of the same dimension as its input, called a *gradient*. For every element of the gradient, the sign tells you which direction to move that coordinate. Following the gradient to maximize a function is called *gradient ascent*; if you're minimizing, the technique is called *gradient descent*.

In this case, it may help to imagine the function as a contoured surface. At any point, the gradient points to the steepest slope of the surface.

To use gradient ascent, you must have a formula for the derivative of the function you're trying to maximize. Most simple algebraic functions have a known derivative; you can look them up in any calculus textbook. If you define a complicated function by chaining many simple functions together, a formula known as the *chain rule* describes how to calculate the derivative of the complicated function. Libraries like TensorFlow and Theano take advantage of the chain rule to automatically calculate the derivative of complicated functions. If you define a complicated function in Keras, you don't need to figure out the formula for the gradient yourself: Keras will hand off the work to TensorFlow or Theano.

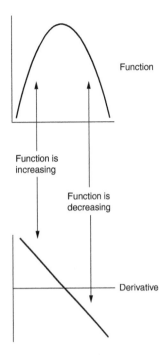

Figure A.2   A function and its derivative. Where the derivative is positive, the function is increasing. Where the derivative is negative, the function is decreasing. When the derivative is exactly zero, the function is at a local minimum or maximum. With this logic, you can use the derivative to find local minima or maxima.

appendix B
The backpropagation algorithm

Chapter 5 introduced sequential neural networks and feed-forward networks in particular. We briefly talked about the *backpropagation algorithm*, which is used to train neural networks. This appendix explains in a bit more detail how to arrive at the gradients and parameter updates that we simply stated and used in chapter 5.

We'll first derive the backpropagation algorithm for feed-forward neural networks and then discuss how to extend the algorithm to more-general sequential and nonsequential networks. Before going deeper into the math, let's define our setup and introduce notation that will help along the way.

A bit of notation

In this section, you'll work with a feed-forward neural network with l layers. Each of the l layers has a sigmoid activation function. Weights of the ith layer are referred to as W^i, and bias terms by b^i. You use x to denote a mini-batch of size k of input data to the network, and y to indicate the output of it. It's safe to think of both x and y as vectors here, but all operations carry over to mini-batches. Moreover, we introduce the following notation:

- We indicate the output of the ith layer with activation y^{i+1}; that is, $y^{i+1} = \sigma(W^i y^i + b^i)$. Note that y^{i+1} is also the *input* to layer $i+1$.
- We indicate the output of the ith dense layer without activation as z^i; that is, $z^i = W^i \cdot y^i + b^i$.
- Introducing this convenient way of writing intermediate output, you can now write $z^i = W^i \cdot y^i + b^i$ and $y^{i+1} = \sigma(z^i)$. Note that with this notation, you could also write the output as $y = y^l$ and the input as $x = y^0$, but we won't use this notation in what follows.
- As a last piece of notation, we'll sometimes write $f^i(y^i)$ for $\sigma(W^i y^i + b^i)$.

318

The backpropagation algorithm for feed-forward networks

Following the preceding conventions, the forward pass for the ith layer of your neural network can now be written as follows:

$$y^{i+1} = \sigma \left(W^i y^i + b^i \right) = f^i \circ y^i$$

You can use this definition recursively for each layer to write your predictions like this:

$$y = f^n \circ \cdots \circ f^1(x)$$

Because you compute your loss function *Loss* from predictions y and labels $\hat{y}$, you can split the loss function in a similar fashion:

$$\text{Loss}\,(y, \hat{y}) = \text{Loss} \circ f^n \circ \cdots \circ f^1(x)$$

Computing and using the derivative of the loss function as shown here is done by a smart application of the *chain rule* for functions, a fundamental result from multivariable calculus. Directly applying the chain rule to the preceding formula yields this:

$$\frac{d\text{Loss}}{dx} = \frac{d\text{Loss}}{df^n} \cdot \frac{df^n}{df^{n-1}} \cdots \frac{df^2}{df^1} \cdot \frac{df^1}{dx}$$

Now, you define the *delta* of the ith layer as follows:

$$\Delta^i = \frac{d\text{Loss}}{df^n} \cdots \frac{df^{i+1}}{df^i}$$

Then you can express deltas in a similar fashion to the previous forward pass, which you call the *backward pass*—namely, by the following relationship:

$$\Delta^i = \Delta^{i+1} \frac{df^{i+1}}{df^i}$$

Note that for deltas, the indices go down, as you pass backward through the computation. Formally, computing the backward pass is structurally equivalent to the simple forward pass. You'll now proceed to explicitly computing the actual derivatives involved. Derivatives of both sigmoid and affine linear functions with respect to their input are quickly derived:

$$\sigma'(x) = \frac{d\sigma}{dx} = \sigma(x)\left(1 - \sigma(x)\right)$$

$$\frac{d\,(Wx + b)}{dx} = W$$

Using these last two equations, you can now write down how to propagate back the error term Δ^{i+1} of the $(i+1)$th layer to the ith layer:

$$\Delta^i = \left(W^i\right)^\top \cdot \left(\Delta^{i+1} \odot \sigma'\left(z^i\right)\right)$$

In this formula, the superscript T denotes matrix transposition. The $\odot$, or *Hadamard product*, indicates element-wise multiplication of the two vectors. The preceding computation splits into two parts, one for the dense layer and one for the activation:

$$\Delta^\sigma = \Delta^{i+1} \odot \sigma'\left(z^i\right)$$

$$\Delta^i = \left(W^i\right)^\top \cdot \Delta^\sigma$$

The last step is to compute the gradients of your parameters W^i and b^i for every layer. Now that you have Δ^i readily computed, you can immediately read off parameter gradients from there:

$$\Delta W^i = \frac{d\text{Loss}}{dW^i} = \Delta^i \cdot \left(y^i\right)^\top$$

$$\Delta b^i = \frac{d\text{Loss}}{db^i} = \Delta^i$$

With these error terms, you can update your neural network parameters as you wish, meaning with any optimizer or update rule you like.

Backpropagation for sequential neural networks

In general, sequential neural networks can have more-interesting layers than what we've discussed so far. For instance, you could be concerned with convolutional layers, as described in chapter 6, or other activation functions, such as the softmax activation discussed in chapter 6 as well. Regardless of the actual layers in a sequential network, backpropagation follows the same general outline. If g^i denotes the forward pass without activation, and Act^i denotes the respective activation function, propagating Δ^{i+1} to the ith layer requires you to compute the following transition:

$$\Delta^i = \frac{d\text{Act}^i}{dg^i}\left(z^i\right)\frac{dg^i}{dz^i}\left(y^i\right)\Delta^{i+1}$$

You need to compute the derivative of the activation function evaluated at the intermediate output z^i and the derivative of the layer function g^i with respect to the input of the ith layer. Knowing all the deltas, you can usually quickly deduce gradients for all parameters involved in the layer, just as you did for weights and bias terms in the feedforward layer. Seen this way, each layer knows how to pass data forward and propagate an error backward, without explicitly knowing anything about the structure of the surrounding layers.

Backpropagation for neural networks in general

In this book, we're concerned solely with sequential neural networks, but it's still interesting to discuss what happens when you move away from the sequentiality constraint. In a nonsequential network, a layer has multiple outputs, multiple inputs, or both.

Let's say a layer has m outputs. A prototypical example might be to split up a vector into m parts. Locally for this layer, the forward pass can be split into k *separate* functions. On the backward pass, the derivative of each of these functions can be computed separately as well, and each derivative contributes equally to the delta that's being passed on to the previous layer.

In the situation that we have to deal with, n inputs and one output, the situation is somewhat reversed. The forward pass is computed from n input components by means of a single function that outputs a single value. On the backward pass, you receive one delta from the next layer and have to compute n output deltas to pass on to each one of the incoming n layers. Those derivatives can be computed independently of each other, evaluated at each of the respective inputs.

The general case of n inputs and m outputs works by combining the two previous steps. Each neural network, no matter how complicated the setup or how many layers in total, locally looks like this.

Computational challenges with backpropagation

You could argue that backpropagation is just a simple application of the chain rule to a specific class of machine-learning algorithms. Although on a theoretical level it may be seen like this, in practice there's a lot to consider when implementing backpropagation.

Most notably, to compute deltas and gradient updates for any layer, you have to have the respective inputs of the forward pass ready for evaluation. If you simply discard results from the forward pass, you have to recompute them on the backward pass. Thus, you'd do well by caching those values in an efficient way. In your implementation from scratch in chapter 5, each layer persisted its own state, for input and output data, as well as for input and output deltas. Building networks that rely on processing massive amounts of data, you should make sure to have an implementation in place that's both computationally efficient and has a low memory footprint.

Another related, interesting consideration is that of reusing intermediate values. For instance, we've argued that in the simple case of a feed-forward network, we can either see affine linear transformation and sigmoid activation as a unit or split them into two layers. The output of the affine linear transformation is needed to compute the backward pass of the activation function, so you should keep that intermediate information from the forward pass. On the other hand, because the sigmoid function doesn't have parameters, you compute the backward pass in one go:

$$\Delta^i = \left(W^i\right)^\top \left(\Delta^{i+1} \odot \sigma'\left(z^i\right)\right)$$

This might computationally be more efficient than doing it in two steps. Automatically detecting which operations can be carried out together can bring a lot of speed gains. In more-complicated situations (such as that of recurrent neural networks, in which a layer will essentially compute a *loop* with inputs from the last step), managing intermediate state becomes even more important.

appendix C
Go programs and servers

This appendix covers various ways to play Go offline and online. First, we show you how to install and play against two Go programs locally—GNU Go and Pachi. Second, we'll point you to a few popular Go servers on which you can find human and AI opponents of various strengths.

Go programs

Let's start with installing Go programs on your computer. We'll introduce you to two classic, free programs that have been around for many years. Both *GNU Go* and *Pachi* use classic game AI methods we partly covered in chapter 4. We introduce these tools not to discuss their methodology, but rather to have two opponents that you can use locally for tests—and to play against them for fun.

As with most other Go programs out there, Pachi and GNU Go can speak the Go Text Protocol (GTP) that we introduced in chapter 8. Both programs can be run in different ways that prove useful for us:

- You can run them from the command line and play games by exchanging GTP commands. This mode is what you use in chapter 8 to let your own bots play against GNU Go and Pachi.
- Both programs can be installed to use *GTP frontends*, graphical user interfaces that make it much more fun to play these Go engines as humans.

GNU Go

GNU Go was developed in 1989 and is one of the oldest Go engines that's still around. The latest release was in 2009. Although there has been little recent development, GNU Go remains a popular AI opponent for beginners on many Go servers. In addition, it's one of the strongest Go engines based on handcrafted rules; this provides a nice contrast to MCTS and deep-learning bots. You can download and install GNU Go from www.gnu.org/software/gnugo/download.html for Windows, Linux, and macOS. This page includes instructions for installing GNU Go as a command-line interface (CLI) tool and links to various graphical interfaces. To install the CLI tool, you need to download the latest GNU Go binaries from http://ftp.gnu.org/gnu/gnugo/, unpack the respective tarball, and follow the instructions for your platform in the INSTALL and README files included in the download. For graphical interfaces, we recommend installing JagoClient for Windows and Linux from www.rene-grothmann.de/jago/ and FreeGoban for macOS from http://sente.ch/software/goban/freegoban.html. To test your installation, you can run the following:

```
gnugo --mode gtp
```

This starts GNU Go in GTP mode. The program will start a new game on a 19 × 19 board and accepts input from the command line. For instance, you could ask GNU Go to generate a white move by typing genmove white and pressing Enter. This returns an = symbol to signify a valid command, followed by coordinates of a move. For instance, the response could be = C3. In chapter 8, you use GNU Go in GTP mode as an opponent for your own deep-learning bots.

When you opt to install a graphical interface, you can start playing a game against GNU Go right away and test your own skills.

Pachi

You can find Pachi, a program considerably stronger than GNU Go overall, for download at http://pachi.or.cz/. Also, Pachi's source code and detailed installation instructions can be found on GitHub at https://github.com/pasky/pachi. To test Pachi, run pachi on the command line and type genmove black to let it generate a black move on a 9 × 9 board for you.

Go servers

Playing against Go programs on your computer can be fun and useful, but online Go servers provide a much richer and stronger pool of human and AI opponents. Humans and bots can register accounts on these platforms and play ranked games to improve their level of game play and ultimately their ratings. For humans, this provides a more competitive and interactive playing field, as well as the ultimate test for your bot to be exposed to players around the world. You can access an extensive list of Go servers at Sensei's Library, https://senseis.xmp.net/?GoServers. We present three servers with *English clients* here. This is a biased list, because by far the largest Go servers are Chinese, Korean, or Japanese and don't come with English-language support. Because this book is written in English, we want to give you access to Go servers that you can navigate in this language.

OGS

Online Go Server (OGS) is a beautifully designed, web-based Go platform that you can find at https://online-go.com/. OGS is the Go server we use to demonstrate how to connect a bot in chapter 8 and appendix E. OGS is feature rich, updated frequently, has an active group of administrators, and is one of the most popular Go servers in the Western hemisphere. On top of that, we like it a lot.

IGS

Internet Go Server (IGS), available at http://pandanet-igs.com/communities/pandanet, was created in 1992 and is one of the oldest Go servers out there. It continues to be popular and has gotten a facelift with a new interface in 2013. It's one of the few Go servers with a native Mac client. IGS is among the more competitive Go servers and has a global user base.

Tygem

Based in Korea, *Tygem* is probably the Go server with the largest user base of the three presented here; no matter what time of day you log on, you'll find thousands of players at all levels. It's also competitive. Many of the world's strongest Go professionals play on Tygem (sometimes anonymously). You can find it at www.tygemgo.com.

appendix D
Training and
deploying bots by using
Amazon Web Services

In this appendix, you're going to learn how to use the cloud service Amazon Web Services (AWS) to build and deploy your deep-learning models. Knowing how to use a cloud service and hosting models is a useful skill in general, not only for this Go bot use case. You'll learn the following skills:

- Setting up a virtual server with AWS to train deep-learning models
- Running deep-learning experiments in the cloud
- Deploying a Go bot with a web interface on a server to share with others

Although as of this writing AWS is the largest cloud provider in the world and provides many benefits, we could've chosen many other cloud services for this appendix. Because the big cloud providers largely overlap in terms of their offerings, getting started with one will help you know the others as well.

To get started with AWS, head over to https://aws.amazon.com/ to see the large product range AWS has to offer. Amazon's cloud service gives you access to an almost intimidatingly large number of products, but for this book you'll get pretty far by using just a single service: Amazon Elastic Compute Cloud (EC2). EC2 gives you easy access to virtual servers in the cloud. Depending on your needs, you can equip these servers or instances with various hardware specifications. To train deep neural networks efficiently, you need access to strong GPUs. Although AWS may not always provide the latest generation of GPUs, flexibly buying compute time on a cloud GPU is a good way to get started without investing too much in hardware up front.

The first thing you need to do is register an account with AWS at https://portal .aws.amazon.com/billing/signup; fill out the form shown in figure D.1.

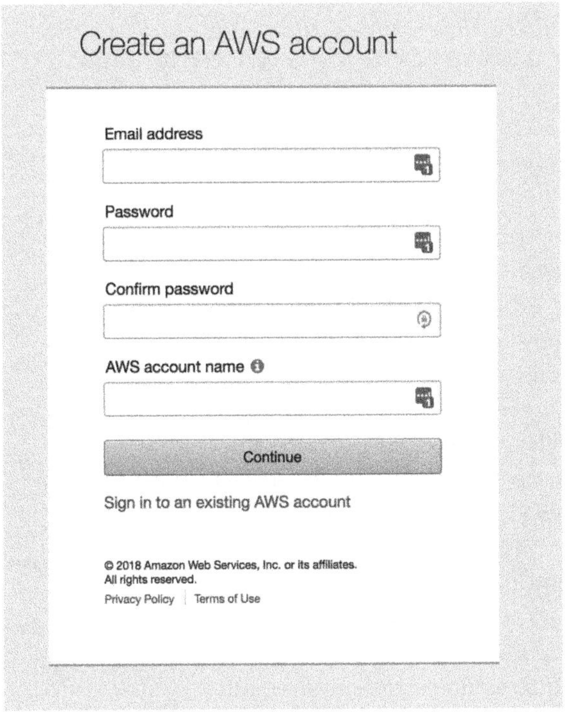

Figure D.1 Signing up for an AWS account

After signing up, at the top right of the page (https://aws.amazon.com/) you should click Sign in to the Console and enter your account credentials. This redirects you to the main dashboard for your account. From the top menu bar, click Services, which opens a panel displaying the AWS core products. Click the EC2 option in the Compute category, as shown in figure D.2.

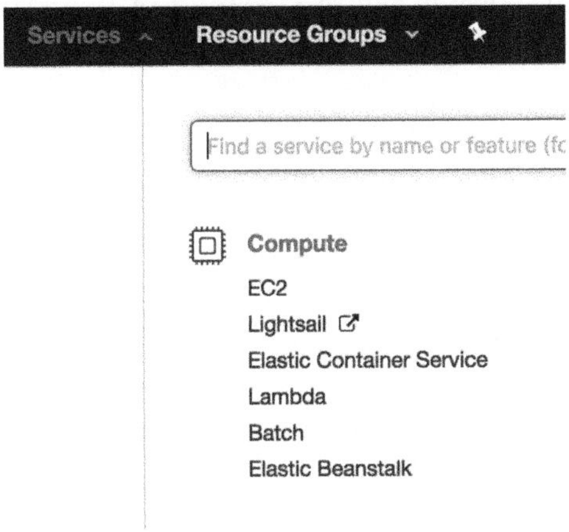

Figure D.2 Selecting the Elastic Cloud Compute (EC2) service from the Services menu

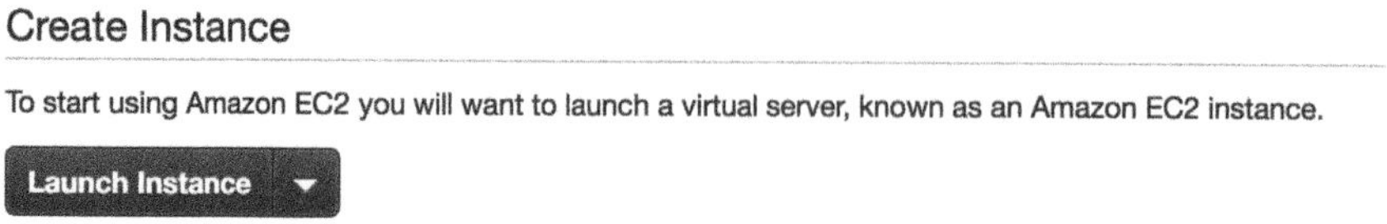

Create Instance

To start using Amazon EC2 you will want to launch a virtual server, known as an Amazon EC2 instance.

Launch Instance ▼

Figure D.3   Launching a new AWS instance

This puts you into the EC2 dashboard, which gives you an overview of your currently running instances and their statuses. Given that you just signed up, you should see 0 running instances. To launch a new instance, click the Launch Instance button, as shown in figure D.3.

At this point, you're asked to select an Amazon Machine Image (AMI), which is a blueprint for the software that will be available to you on your launched instance. To get started quickly, you'll choose an AMI that's specifically tailored for deep-learning applications. On the left sidebar, you'll find the AWS Marketplace (see figure D.4), which has a lot of useful third-party AMIs.

In the Marketplace, search for Deep Learning AMI Ubuntu, as shown in figure D.5. As the name suggests, this instance runs on Ubuntu Linux and has many useful components already preinstalled. For instance, on this AMI, you'll find TensorFlow and Keras available, plus all the

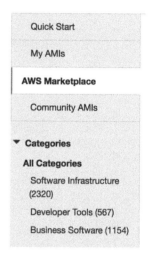

Figure D.4   Selecting the AWS Marketplace

necessary GPU drivers already installed for you. Therefore, when the instance is ready, you can get right into your deep-learning application, instead of spending time and effort installing software.

Choosing this particular AMI is cheap but doesn't come entirely for free. If you want to play with a free instance instead, look for the *free tier eligible* tag. For example, in the Quick Start section shown previously in figure D.4, most of the AMIs shown there you *can* get for free.

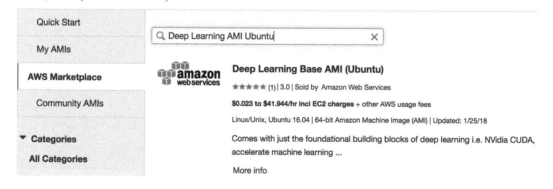

Figure D.5   Choosing an AMI suited for deep learning

Deep Learning Base AMI (Ubuntu)

Deep Learning Base AMI (Ubuntu)

Comes with just the foundational building blocks of deep learning i.e. NVidia CUDA, cuDNN, GPU drivers, and low-level system libraries to scale and accelerate machine learning operations on AWS EC2 instances. The base AMI serves as a clean slate to deploy your customized deep learning set up.

For example, for developers contributing to open ...
More info

View Additional Details in AWS Marketplace

Product Details

Pricing Details

Hourly Fees

Instance Type	Software	EC2	Total
R3 Eight Extra Large	$0.00	$3.201	**$3.201/hr**
M3 Extra Large	$0.00	$0.315	**$0.315/hr**
R4 16 Extra Large	$0.00	$5.122	**$5.122/hr**
M4 Extra Large	$0.00	$0.24	**$0.24/hr**
Graphics Two Extra Large	$0.00	$0.772	**$0.772/hr**
C3 Quadruple Extra Large	$0.00	$1.032	**$1.032/hr**
High I/O Quadruple Extra Large	$0.00	$1.488	**$1.488/hr**

Figure D.6   **Pricing for your deep-learning AMI, depending on the instance you choose**

After clicking Select for the AMI of your choice, a tab opens that shows you the pricing for this AMI, depending on which instance type you choose; see figure D.6.

Continuing, you can now choose your instance type. In figure D.7, you see all instance types optimized for GPU performance. Selecting p2.xlarge is a good option to get started, but keep in mind that all GPU instances are comparatively expensive. If you first want to get a feel for AWS and familiarize yourself with the features presented here, go for an inexpensive t2.small instance first. If you're interested only in deploying and hosting models, a t2.small instance will be sufficient anyway; it's only the model training that requires more-expensive GPU instances.

After you've chosen an instance type, you could directly click the Review and Launch button in the lower right to immediately launch the instance. But because you still need to configure a few things, you'll instead opt for Next: Configure Instance Details. Steps 3 to 5 in the dialog box that follows can be safely skipped for now, but step 6 (Configure Security Group) requires some attention. A *security group* on AWS

	GPU graphics	g3.4xlarge	16	122	EBS only
	GPU graphics	g3.8xlarge	32	244	EBS only
	GPU graphics	g3.16xlarge	64	488	EBS only
	GPU instances	g2.2xlarge	8	15	1 x 60 (SSD)
	GPU instances	g2.8xlarge	32	60	2 x 120 (SSD)
	GPU compute	p2.xlarge	4	61	EBS only
	GPU compute	p2.8xlarge	32	488	EBS only
	GPU compute	p2.16xlarge	64	732	EBS only

Figure D.7   **Selecting the right instance type for your needs**

specifies access rights to the instance by defining *rules.* You want to grant the following access rights:

- Primarily, you want to access your instance by logging in through SSH. The SSH port 22 on the instance should already by open (this is the only specified rule on new instances), but you need to restrict access and allow connections from only your local machine. You do this for security reasons, so that nobody else can access your AWS instance; only your IP is granted access. This can be achieved by selecting My IP under Source.
- Because you also want to deploy a web application, and later even a bot that connects to other Go servers, you'll also have to open HTTP port 80. You do so by first clicking Add Rule and selecting HTTP as the type. This will automatically select port 80 for you. Because you want to allow people to connect to your bot from anywhere, you should select Anywhere as the Source.
- The HTTP Go bot from chapter 8 runs on port 5000, so you should open this port as well. In production scenarios, you'd normally deploy a suitable web server listening on port 80 (which you configured in the preceding step); this internally redirects traffic to port 5000. To keep things simple, you trade security for convenience and open port 5000 directly. You can do this by adding another rule and selecting Custom TCP Rule as the type and 5000 as the port range. As for the HTTP port, you set the source to Anywhere. This will prompt a security warning, which you'll ignore because you're not dealing with any sensitive or proprietary data or application.

If you configured the access rules as we just described, your settings should look like those in figure D.8.

After completing security settings, you can click Review and Launch and then Launch. This opens a window that will ask you to create a new key pair or select an

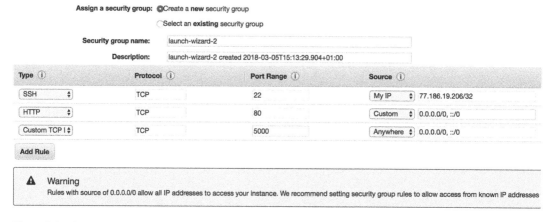

Figure D.8   Configuring security groups for your AWS instance

existing one. You need to select Create a New Pair from the drop-down menu. The only thing you need to do is select a *key pair name* and then download the *secret* key by clicking Download Key Pair. The downloaded key will have the name you've given it, with a .pem file signature. Make sure to store this private key in a secure location. The public key for your private key is managed by AWS and will be put on the instance you're about to launch. With the private key, you can then connect to the instance. After you've created a key, you can reuse it in the future by selecting Choose an Existing Key Pair. In figure D.9, you see how we created a key pair called maxpumperla_aws.pem.

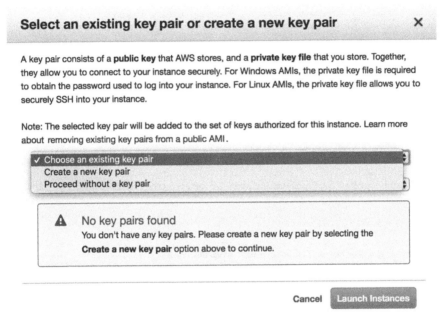

Figure D.9 Creating a new key pair to access your AWS instance

This was the final step, and you can now launch your instance by clicking Launch Instance. You'll see an overview called Launch Status, and you can proceed by selecting View Instances in the lower right. This puts you back into the EC2 main dashboard from which you started (by selecting Launch Instances). You should see your instance listed there. After waiting for a bit, you should see that the instance state is "running" and see a green dot next to the state. This means your instance is ready, and you can now connect to it. You do so by first selecting the check box to the left of the instance, which activates the Connect button on top. Clicking this button opens a window that looks like the one shown in figure D.10.

This window contains a lot of useful information for connecting to your instance, so read it carefully. In particular, it gives you instructions on how to connect to your instance with ssh. If you open a terminal and then copy and paste the ssh command

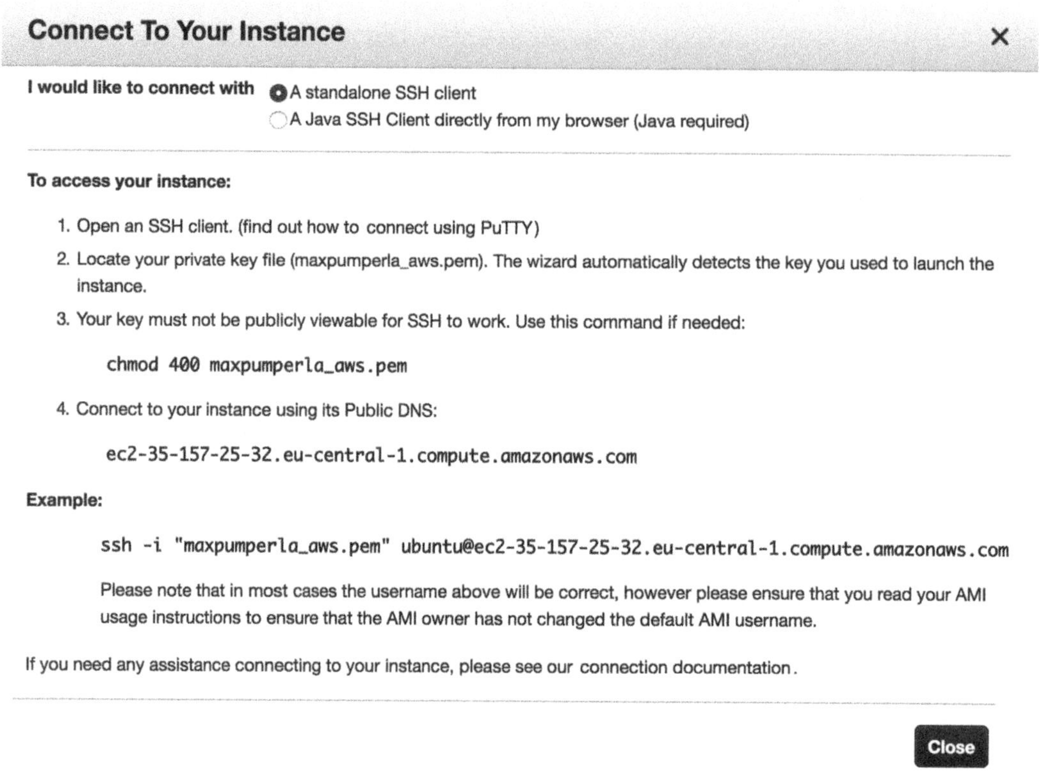

Connect To Your Instance   ✕

I would like to connect with ⦿ A standalone SSH client
                                  ◯ A Java SSH Client directly from my browser (Java required)

To access your instance:

1. Open an SSH client. (find out how to connect using PuTTY)

2. Locate your private key file (maxpumperla_aws.pem). The wizard automatically detects the key you used to launch the instance.

3. Your key must not be publicly viewable for SSH to work. Use this command if needed:

    chmod 400 maxpumperla_aws.pem

4. Connect to your instance using its Public DNS:

    ec2-35-157-25-32.eu-central-1.compute.amazonaws.com

Example:

    ssh -i "maxpumperla_aws.pem" ubuntu@ec2-35-157-25-32.eu-central-1.compute.amazonaws.com

Please note that in most cases the username above will be correct, however please ensure that you read your AMI usage instructions to ensure that the AMI owner has not changed the default AMI username.

If you need any assistance connecting to your instance, please see our connection documentation.

[Close]

Figure D.10   Creating a new key pair to access your AWS instance

listed under Example, you should establish a connection to your AWS instance. This command is as follows:

```
ssh -i "<full-path-to-secret-key-pem>" <username>@<public-dns-of-your-instance>
```

This is a long command that can be a little inconvenient to work with, especially when you're handling many instances or SSH connections to other machines as well. To make life easier, we're going to work with an SSH configuration file. In UNIX environments, this configuration file is usually stored at ~/.ssh/config. On other systems, this path may vary. Create this file and the .ssh folder, if necessary, and put the following content into this file:

```
Host aws
  HostName <public-dns-of-your-instance>
  User ubuntu
  Port 22
  IdentityFile <full-path-to-secret-key-pem>
```

Having stored this file, you can now connect to your instance by typing `ssh aws` into your terminal. When you first connect, you're asked whether you want to connect. Type yes and submit this command by pressing Enter. Your key will be added permanently to the instance (which you can check by running `cat ~/.ssh/authorized_keys` to return a secure hash of your key pair), and you won't be asked again.

The first time you successfully log into the instance of the Deep Learning AMI Ubuntu AMI (in case you went with this one), you'll be offered a few Python environments to choose from. An option that gives you a full Keras and TensorFlow installation for Python 3.6. is `source activate tensorflow_p36`, or `source activate tensorflow_p27` if you prefer to go with Python 2.7. For the rest of this appendix, we assume you skip this and work with the basic Python version already provided on this instance.

Before you proceed to running applications on your instance, let's quickly discuss how to terminate an instance. This is important to know, because if you forget to shut down an expensive instance, you can easily end up with a few hundred dollars of costs per month. To terminate an instance, you select it (by clicking the check box next to it, as you did before) and then click the Actions button at the top of the page, followed by Instance State and Terminate. Terminating an instance deletes it, including everything you stored on it. Make sure to copy everything you need (for instance, the model you trained) before termination (we'll show you how in just a bit). Another option is to Stop the instance, which allows you to Start it at a later point. Note, however, that depending on the storage that your instance is equipped with, this might still lead to data loss. You'll be prompted with a warning in this situation.

Model training on AWS

Running a deep-learning model on AWS works the same way as running it locally, after you have everything in place. You first need to make sure you have all the code and data you need on the instance. An easy way to do that is by copying it there in a secure way by using `scp`. For example, from your local machine, you can run the following commands to compute an end-to-end example:

Copy code from local to your remote AWS instance.

```
git clone https://github.com/maxpumperla/deep_learning_and_the_game_of_go
cd deep_learning_and_the_game_of_go
scp -r ./code aws:~/code
ssh aws                          ←——— Log into the instance with ssh.
cd ~/code
python setup.py develop          ←——— Install your dlgo Python library.
cd examples
python end_to_end.py             ←——— Run an end-to-end example.
```

In this example, we assume you start fresh by cloning our GitHub repository first. In practice, you'll have done this already and want to build your own experiments instead. You do this by creating the deep neural networks you want to train and running the examples you want. The example end_to_end.py we just presented will produce a serialized deep-learning bot in the following path relative to the examples folder: ../agents/deep_bot.h5. After the example runs, you can either leave the model there (for example, to host it or continue working on it) or retrieve it from the AWS instance and copy it back to your machine. For instance, from a terminal on your local machine, you can copy a bot called deep_bot.h5 from AWS to local as follows:

```
cd deep_learning_and_the_game_of_go/code
scp aws:~/code/agents/deep_bot.h5 ./agents
```

This makes for a relatively lean model-training workflow that we can summarize as follows:

1 Set up and test your deep-learning experiments locally by using the dlgo framework.
2 Securely copy the changes you made to your AWS instance.
3 Log into the remote machine and start your experiment.
4 After training finishes, evaluate your results, adapt your experiment, and start a new experimentation cycle at 1.
5 If you wish, copy your trained model to your local machine for future use or process it otherwise.

Hosting a bot on AWS over HTTP

Chapter 8 showed you how to serve a bot over HTTP so you and your friends can play against it through a convenient web interface. The drawback was that you simply started a Python web server locally on your machine. So for others to test your bot, they'd have to have direct access to your computer. By deploying this web application on AWS and opening the necessary ports (as you did when setting up the instance), you can share your bot with others by sharing a URL.

Running your HTTP frontend works the same way as before. All you need to do is the following:

```
ssh aws
cd ~/code
python web_demo.py \
  --bind-address 0.0.0.0 \
  --pg-agent agents/9x9_from_nothing/round_007.hdf5 \
  --predict-agent  agents/betago.hdf5
```

This hosts a playable demo of your bot on AWS and makes it available under the following address:

```
http://<public-dns-of-your-instance>:5000/static/play_predict_19.html
```

That's it! In appendix E, we'll go one step further and show you how to use the AWS basics presented here to deploy a full-blown bot connecting to the Online Go Server (OGS) using the Go Text Protocol (GTP).

appendix E
Submitting a bot
to the Online Go Server

In this appendix, you'll learn how to deploy a bot to the popular Online Go Server. To do so, you'll use your bot framework from the first eight chapters of this book to deploy a bot on the cloud-provided Amazon Web Services (AWS) that communicates through the Go Text Protocol (GTP). Make sure to read the first eight chapters to understand the basics of this framework and read up on appendix D for AWS basics.

Registering and activating your bot at OGS

The *Online Go Server* (OGS) is a popular platform on which you can play Go against other human players and bots. Appendix C showed you a few other Go servers, but we've picked OGS for this appendix to demonstrate how to deploy bots. OGS is a modern web-based platform that you can explore at https://online-go.com. To register with OGS, you have to sign up at https://online-go.com/register. If you want to deploy a bot with OGS, you need to create *two* accounts:

1 Register an account for yourself as a human player. Pick an available username, password, and optionally enter your email address. You can also register via Google, Facebook, or Twitter. We'll refer to this account as <human>.

2 Go back to the registration one more time and register another account. This will act as your bot account, so give it an appropriate name, signifying it as a bot. We'll call this account <bot> in what follows.

At this point, all you have is two regular accounts. What you want to achieve is to make the second account a *bot account* that's owned and managed by the *user account*. To make this happen, you first need to sign in with your human account on OGS and find an OGS moderator that can activate your bot account. In the top left, next to the OGS logo, you can open the menu and search for users by name. OGS moderators `crocrobot` and `anoek` offered help with the registration process for this book. If you search for either one of those names and then click the account name in the search results, a pop-up box like the one shown in figure E.1 should open.

In this box, click Message to get in touch with the moderator. A message box should open in the lower right. You'll have to tell the moderator that you want to activate a bot account for <bot> and that this bot belongs to the human account <human> (the one you're currently logged into). Usually, the OGS moderators get back to you within 24 hours, but you might have to be a little patient. You find a moderator under the Chat option in the top OGS menu; every user with a hammer symbol next to their name is an OGS moderator. If a moderator is on vacation

Figure E.1   Contacting an OGS moderator to activate your bot account

or otherwise busy, you might find someone else who can help you out.

If you have trouble getting in touch with a moderator directly, you can also try creating a message in the OGS forum (https://forums.online-go.com), in the OGS Development section. Remember that the moderators are all volunteers who help out in their spare time, so be patient!

After you hear back from the moderator you got in touch with, you can log into your <bot> account. Click the menu symbol at the top left of the OGS page and select Profile to view the profile page of your bot. If all went well, your <bot> account should be listed as Artificial Intelligence and should have an Administrator, namely your <human> account. In short, your bot profile should look similar to what you see for the `BetagoBot` account in figure E.2, which is administrated by Max's human account `DoubleGotePanda`.

Next, log out of your <bot> account and back into your <human> account. You need to do this to generate an API key for your bot, which can be done only through its human administrator account. Once logged into <human>, visit the profile page of <bot> (for instance, by searching for and clicking <bot>). Scrolling down a little, you'll find a Bot Controls box containing a Generate API Key button. Click this button to generate your API key and then click Save to store it. For the rest of this appendix, we'll assume your API key is called <api-key>.

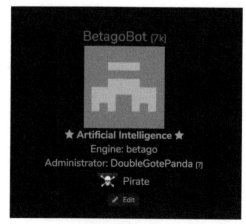

Figure E.2   Checking the profile page of your bot to see it has been activated

Now that you've set up everything on OGS, you'll proceed to using your bot name and API key to connect a GTP bot to OGS.

Testing your OGS bot locally

In chapter 8, you developed a bot that can understand and emit GTP commands. You now also have a bot account on OGS. The missing link to connect the two is a tool called gtp2ogs, which takes your bot name and API key and establishes a connection between the machine that has your bot and OGS. gtp2ogs is an open source library built in Node.js and is available in the official OGS GitHub repository under https://github.com/online-go/gtp2ogs. You don't need to download or install this tool, because we've provided you with a copy of it in our GitHub repository already. In your local copy of http://mng.bz/gYPe, you should see a file called gtp2ogs.js and a JSON file called package.json. The latter you use to install dependencies; the former is the tool itself.

When deploying a bot to OGS, you want this bot to be available for everyone to play—for a long time. This deployment task is that of a *long-running process*. For that reason, it makes sense to serve the bot from a (remote) server. You'll do precisely this in the next section, but you can first quickly test if everything works using your local machine. To that end, make sure you have both Node.js and its package manager (npm) installed on your system. On most systems, you can get both from the package manager of your choice (for example, running `brew install node npm` on a Mac or `sudo apt-get install npm nodejs-legacy` on Ubuntu), but you can also download and install these tools from https://nodejs.org/en/download/.

Next, you need to put the run_gtp.py Python script that you find at the top level in our GitHub repo on your *system path*. In Unix environments, you can do this from the command line by executing the following command:

```
export PATH=/path/to/deep_learning_and_the_game_of_go/code:$PATH
```

This puts run_gtp.py on your path, so that you can call it from anywhere on the command line. In particular, it'll be available to gtp2ogs, which will spawn a new bot using run_gtp.py whenever a new game is requested for your bot on OGS. Now all that is left is to install the necessary Node.js packages and run the application. You'll use the Node.js package *forever* to ensure that the application stays up and restarts should it fail at some point:

```
cd deep_learning_and_the_game_of_go/code
npm install

forever start gtp2ogs.js \
    --username <bot> \
    --apikey <api-key> \
    --hidden \
    --persist \
    --boardsize 19 \
    --debug -- run_gtp.py
```

Let's break down that command line:

- `--username` and `--apikey` specify how to connect to the server.
- `--hidden` keeps your bot out of the public bot lists, which will give you a chance to test everything before other players start challenging your bot.
- `--persist` keeps your bot running between moves (otherwise, gtp2ogs will restart your bot every time it needs to make a move).
- `--boardsize 19` limits your bot to accept 19 x 19 games; if you trained your bot to play 9 × 9 (or some other size), use that instead.
- `--debug` prints out extra logging so you can see what your bot is doing.

When your bot is running, head over to OGS, log into your <human> account, and click on the left menu. Type your bot's name into the search box, click its name, and then click the Challenge button. Then you can start a match against your bot and start playing.

If you can select your bot, everything likely worked well, and you can now play your first game against your own creation. After you tested connecting your bot successfully, stop the Node.js application running your bot by typing `forever stopall`.

Deploying your OGS bot on AWS

Next, we'll show you how to deploy your bot on AWS for free, so you and many other players around the world can play against it anytime (without having to run a Node.js application on your local computer).

For this part, we're going to assume that you followed appendix D and have your SSH config configured so that you can access your AWS instance with `ssh aws`. The instance you use can be limited, because you don't need much compute power to generate predictions from an already trained deep-learning model. In fact, you could resort to using one of the free-tier-eligible instances on AWS, like t2.micro. If you follow appendix D literally and choose a Deep Learning AMI for Ubuntu running on a t2.small, that won't be entirely free but will cost only a few dollars per month—in case you want to keep your bot running on OGS.

In our GitHub repository, you'll find a script called run_gtp_aws.py, which appears in listing E.1 below. The first line, starting with #!, tells the Node.js process which Python installation to use to run your bot. The base Python installation on your AWS instance should be something like /usr/bin/python, which you can check by typing `which python` in the terminal. Make sure this first line points to the Python version you've been using to install dlgo.

Listing E.1 run_gtp_aws.py to run a bot on AWS that connects against OGS

```
#!/usr/bin/python          ⟵    It's important to make sure this matches the
from dlgo.gtp import GTPFrontend    output of "which python" on your instance.
from dlgo.agent.predict import load_prediction_agent
from dlgo.agent import termination
import h5py
```

```
model_file = h5py.File("agents/betago.hdf5", "r")
agent = load_prediction_agent(model_file)
strategy = termination.get("opponent_passes")
termination_agent = termination.TerminationAgent(agent, strategy)

frontend = GTPFrontend(termination_agent)
frontend.run()
```

This script loads an agent from the file, initializes a termination strategy, and runs an instance of a `GTPFrontend` as defined in chapter 8. The agent and termination strategy chosen are for illustration purposes. You can modify both to your needs and use your own trained models and strategies instead. But to get started and familiarize yourself with the process of submitting a bot, you can leave the script as is for now.

Next, you need to make sure you have everything installed on your AWS instance to run the bot. Let's start fresh, clone your GitHub repo locally, copy it to the AWS instance, log into it, and install the dlgo package:

```
git clone https://github.com/maxpumperla/deep_learning_and_the_game_of_go
cd deep_learning_and_the_game_of_go
scp -r ./code aws:~/code
ssh aws
cd ~/code
python setup.py develop
```

This is essentially the same set of steps you carried out to run an end-to-end example in appendix D. To run *forever* and *gtp2ogs*, you also need to make sure you have Node.js and npm available. After installing these programs on AWS by using apt, you can install gtp2ogs the same way you did locally:

```
sudo apt install npm
sudo apt install nodejs-legacy
npm install
sudo npm install forever -g
```

The final step is to run the GTP bot by using gtp2ogs. You export your current working directory to the system path and use run_gtp_aws.py as the bot runner this time:

```
PATH=/home/ubuntu/code:$PATH forever start gtp2ogs.js \
  --username <bot> \
  --apikey <api-key> \
  --persist \
  --boardsize 19 \
  --debug -- run_gtp_aws.py > log 2>&1 &
```

Note that you redirect standard output and error messages into a log file called *log* and start the program as a background process with `&`. This way, your command line on the instance isn't cluttered by server logs, and you can continue working on that machine. As in your local test of the OGS bot, you should now be able to connect to

OGS and play a game against your bot. In case something breaks or doesn't work as expected, you can check your latest bot logs by inspecting `tail log`.

That's all there is to it. Although it takes some time to set up this pipeline (in particular, creating an AWS instance and setting up the two OGS accounts), after the basics are out of the way, deploying a bot is fairly straightforward. When you've developed a new bot locally and want to deploy it, all you do is the following:

```
scp -r ./code aws:~/code
ssh aws
cd ~/code
PATH=/home/ubuntu/code:$PATH node gtp2ogs.js \
  --username <bot> \
  --apikey <api-key> \
  --persist \
  --boardsize 19 \
  --debug -- run_gtp_aws.py > log 2>&1 &
```

Now that your bot is running without the `--hidden` option, it's open to challenges from the whole server. To find your bot, log into your <human> account, and click Play on the main menu. In the Quick Match Finder, click Computer to select a bot to play against. The name of your bot, <bot>, should show up in the drop-down menu as AI Player. In figure E.3, you can see the BetagoBot that Max and Kevin developed. Right now, you find just a handful of bots on OGS—maybe you can add an interesting bot? This completes the appendix. You can now deploy an end-to-end machine-learning pipeline resulting in a playable bot on an online Go platform.

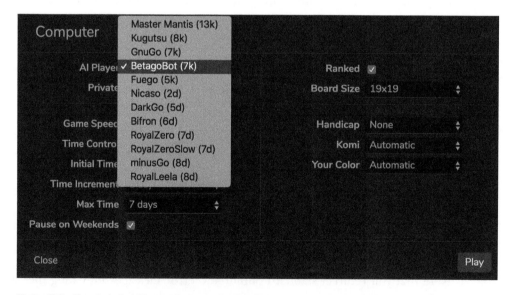

Figure E.3   Your bot should now show up as a Computer opponent in the OGS match finder.

index

Deep Learning with Python
by François Chollet

    ISBN: 9781617294433
    384 pages
    $49.99
    November 2017

Deep Learning with R
by François Chollet
   with J. J. Allaire

    ISBN: 9781617295546
    360 pages
    $49.99
    January 2018

For ordering information go to www.manning.com

MORE TITLES FROM MANNING

Grokking Deep Learning
by Andrew W. Trask

ISBN: 9781617293702
336 pages
$49.99
January 2019

Keras in Motion
by Dan Van Boxel

Course duration: 2h 4m
55 exercises
$49.99

For ordering information go to www.manning.com

The Quick Python Book,
Third Edition
by Naomi Ceder

ISBN: 9781617294037
472 pages
$39.99
May 2018

Natural Language Processing
in Action
Understanding, analyzing, and
generating text with Python
by Hobson Lane, Cole Howard,
  and Hannes Max Hapke

ISBN: 9781617294631
420 pages
$49.99
January 2019

For ordering information go to www.manning.com

MORE TITLES FROM MANNING

Practical Recommender Systems
by Kim Falk

> ISBN: 9781617292705
> 400 pages
> $49.99
> February 2019

Grokking Algorithms
An illustrated guide for programmers
and other curious people

by Aditya Y. Bhargava

> ISBN: 9781617292231
> 256 pages
> $44.99
> May 2016

For ordering information go to www.manning.com